T0213004

Lecture Notes in Computer Science   **9668**

Commenced Publication in 1973
Founding and Former Series Editors:
Gerhard Goos, Juris Hartmanis, and Jan van Leeuwen

Editorial Board

David Hutchison
  Lancaster University, Lancaster, UK
Takeo Kanade
  Carnegie Mellon University, Pittsburgh, PA, USA
Josef Kittler
  University of Surrey, Guildford, UK
Jon M. Kleinberg
  Cornell University, Ithaca, NY, USA
Friedemann Mattern
  ETH Zurich, Zürich, Switzerland
John C. Mitchell
  Stanford University, Stanford, CA, USA
Moni Naor
  Weizmann Institute of Science, Rehovot, Israel
C. Pandu Rangan
  Indian Institute of Technology, Madras, India
Bernhard Steffen
  TU Dortmund University, Dortmund, Germany
Demetri Terzopoulos
  University of California, Los Angeles, CA, USA
Doug Tygar
  University of California, Berkeley, CA, USA
Gerhard Weikum
  Max Planck Institute for Informatics, Saarbrücken, Germany

More information about this series at http://www.springer.com/series/7407

Maria J. Blesa · Christian Blum
Angelo Cangelosi · Vincenzo Cutello
Alessandro Di Nuovo · Mario Pavone
El-Ghazali Talbi (Eds.)

Hybrid Metaheuristics

10th International Workshop, HM 2016
Plymouth, UK, June 8–10, 2016
Proceedings

 Springer

Editors

Maria J. Blesa
Universitat Politècnica de Catalunya
Barcelona
Spain

Christian Blum
IKERBASQUE and University of the
  Basque Country
San Sebastián
Spain

Angelo Cangelosi
Universiy of Plymouth
Plymouth
UK

Vincenzo Cutello
University of Catania
Catania
Italy

Alessandro Di Nuovo
Sheffield Hallam University
Sheffield
UK

Mario Pavone
University of Catania
Catania
Italy

El-Ghazali Talbi
University of Lille 1
Villeneuve d'Ascq
France

ISSN 0302-9743   ISSN 1611-3349   (electronic)
Lecture Notes in Computer Science
ISBN 978-3-319-39635-4   ISBN 978-3-319-39636-1   (eBook)
DOI 10.1007/978-3-319-39636-1

Library of Congress Control Number: 2016939578

LNCS Sublibrary: SL1 – Theoretical Computer Science and General Issues

© Springer International Publishing Switzerland 2016
This work is subject to copyright. All rights are reserved by the Publisher, whether the whole or part of the material is concerned, specifically the rights of translation, reprinting, reuse of illustrations, recitation, broadcasting, reproduction on microfilms or in any other physical way, and transmission or information storage and retrieval, electronic adaptation, computer software, or by similar or dissimilar methodology now known or hereafter developed.
The use of general descriptive names, registered names, trademarks, service marks, etc. in this publication does not imply, even in the absence of a specific statement, that such names are exempt from the relevant protective laws and regulations and therefore free for general use.
The publisher, the authors and the editors are safe to assume that the advice and information in this book are believed to be true and accurate at the date of publication. Neither the publisher nor the authors or the editors give a warranty, express or implied, with respect to the material contained herein or for any errors or omissions that may have been made.

Printed on acid-free paper

This Springer imprint is published by Springer Nature
The registered company is Springer International Publishing AG Switzerland

Preface

The HM workshops are intended to be an international forum for researchers in the area of design, analysis, and experimental evaluation of metaheuristics and their integration with techniques typical of other fields. Metaheuristics, such as simulated annealing, evolutionary algorithms, tabu search, ant colony optimization, scatter search, and iterated local search, are considered state-of-the-art methods for many problems. In recent years, however, it has become evident that the concentration on a sole metaheuristic is rather restrictive. A skilled combination of concepts from different optimization techniques can provide a more efficient behavior and a higher flexibility when dealing with real-world and large-scale problems. Hybrid metaheuristics are such techniques for optimization that combine different metaheuristics or integrate AI/OR techniques into metaheuristics.

The first edition of HM was held in 2004 and, since then, the event has been held regularly. HM 2016 was already the tenth edition of the Workshop on Hybrid Metaheuristics. The preceding workshops were held in Hamburg (2014), Ischia Island (HM 2013), Vienna (HM 2010), Udine (HM 2009), Malaga (HM 2008), Dortmund (HM 2007), Gran Canaria (HM 2006), Barcelona (HM 2005), and Valencia (HM 2004). Except for its first edition, the accepted papers of previous HM workshops were published by Springer in the series *Lecture Notes in Computer Science* (LNCS 3636, LNCS 4030, LNCS 4771, LNCS 5296, LNCS 5818, LNCS 6373, LNCS 7919, LNCS 8457).

HM 2016 continued to be the only three-day event entirely dedicated to the integration of metaheuristics and classic techniques typical of other fields, with the primary aim of providing researchers and scholars with a wide forum for discussing new ideas and new research directions. In addition to learning more about their own research area, the workshop has served to make researchers aware of how their research might contribute and become really fruitful also in other research areas.

As always, this edition confirmed that hybrid metaheuristics are indeed robust and effective, and that several research areas can be put together. Slowly but surely, this process has been promoting productive dialogue among researchers with different expertise and eroding barriers between research areas.

HM 2016 received an overall of 43 submissions from different countries, between regular manuscripts and abstracts, with a total of 16 works accepted (15 full papers and one extended abstract) on the basis of reviews by the Program Committee members and evaluations by the program chairs. There was one additional abstract for oral presentation only. In keeping with the tradition, we had a double-blind peer review process, with four to five expert referees per manuscript, so that not only the originality and overall quality of the papers could be properly evaluated, but also constructive suggestions for improvement could be provided. In light of this, a special thanks is addressed to each member of the Program Committee and external reviewers for devoting their valuable time.

The present selection of manuscripts is of interest to all the researchers working on integrating metaheuristics with other areas for solving both optimization and constraint satisfaction problems. It also represents a sample of current research demonstrating how metaheuristics can be integrated with integer linear programming and other operational research techniques for tackling difficult and relevant problems.

HM 2016 was held in Plymouth, UK, during June 8–10, 2016, and was enriched by three excellent plenary speakers: Carlos A. Coello, Jin- Kao Hao, and Helena Ramalhinho Lourenço. We would like to express our gratitude to them all for having accepted our invitation, and for their participation, which greatly enhanced the quality of the workshop.

Finally, we would like to express our gratitude to everyone that helped us in any way for the success of HM 2016, beginning of course with all the authors who supported the workshop by sending their excellent contributions; and all those who participated in these three days entirely dedicated to science. A special thanks is also addressed to the publicity chair, Antonio Masegosa, for his great job and his valuable support for the success of HM 2016. Without these components, we would not have been able to organize a successful scientific congress.

June 2016   Maria J. Blesa
                                                                 Christian Blum
                                                               Angelo Cangelosi
                                                                Vincenzo Cutello
                                                             Alessandro Di Nuovo
                                                                   Mario Pavone
                                                              El-Ghazali Talbi

Organization

Organizing Committee

Honorary Chair (2016 edition)

Angelo Cangelosi	University of Plymouth, UK

General Chairs

Vincenzo Cutello	University of Catania, Italy
Alessandro Di Nuovo	Sheffield Hallam University, UK

Program Chairs

Christian Blum	IKERBASQUE and University of the Basque Country, Spain
Mario Pavone	University of Catania, Italy
El-Ghazali Talbi	Polytech'Lille, University of Lille 1, France

Publication Chair

Maria J. Blesa	Universitat Politècnica de Catalunya, Spain

Publicity Chair

Antonio D. Masegosa	IKERBASQUE and University of Deusto, Spain

Local Organization

Pietro Consoli	University of Birmingham, UK
Frederico Klein	University of Plymouth, UK
Debora Zanatto	University of Plymouth, UK
Francois Foerster	University of Plymouth, UK
Quentin Houbre	University of Plymouth, UK

HM Steering Committee

Maria J. Blesa	Universitat Politècnica de Catalunya, Spain
Christian Blum	IKERBASQUE and University of the Basque Country, Spain
Luca di Gaspero	University of Udine, Italy
Paola Festa	Università degli Studi di Napoli Federico II, Italy
Günther Raidl	Vienna University of Technology, Austria
Michael Sampels	Université Libre de Bruxelles, Belgium

Keynote Speakers

Carlos A. Coello Coello   CINVESTAV, México
Jin-Kao Hao   University of Angers, France
Helena Ramalhinho   Universitat Pompeu Fabra, Barcelona, Spain
  Lourenço

Program Committee

Andy Adamatzky   University of the West of England, UK
Lyuba Alboul   Sheffield Hallam University, UK
Jason Atkin   University of Nottingham, UK
Peter J. Bentley   University College London, UK
Mauro Birattari   IRIDIA, Université Libre de Bruxelles, Belgium
Jacek Blazewicz   University of Poznan, Poland
Juergen Branke   Warwick Business School, UK
Fabio Caraffini   De Montfort University, UK
Francesco Carrabs   University of Salerno, Italy
Carmine Cerrone   University of Salerno, Italy
Raffaele Cerulli   University of Salerno, Italy
Francisco Chicano   University of Malaga, Spain
Camelia Chira   Technical University of Cluj-Napoca, Romania
Carlos A. Coello Coello   CINVESTAV-IPN, México
Pietro Consoli   University of Birmingham, UK
Xavier Delorme   ENSM-SE, France
Antonio Di Nola   University of Salerno, Italy
Marco Dorigo   IRIDIA, Université Libre de Bruxelles, Belgium
Luca di Gaspero   University of Udine, Italy
Karl Doerner   University of Vienna, Austria
Andreas Ernst   Monash University, Australia
Steven Lawrence Fernandes   Sahyadri College of Engineering and Management,
                                 India
Paola Festa   University of Naples Federico II, Italy
Grazziela Figueredo   University of Nottingham, UK
Carlos M. Fonseca   University of Coimbra, Portugal
Luca Maria Gambardella   University of Lugano, Switzerland
Haroldo Gambini-Santos   Universidade Federal de Ouro Preto, Brazil
Salvatore Greco   University of Catania, Italy
Said Hanafi   University of Valenciennces, France
Julia Handl   University of Manchester, UK
Jin-Kao Hao   University of Angers, France
Colin Johnson   University of Kent, UK
Laetitia Jourdan   University of Lille 1, Inria Lille, France
Graham Kendall   University of Nottingham, UK
Natalio Krasnogor   University of Newcastle, UK

Meng-Hiot Lim	Nanyang Technological University, Singapore
Marco Locatelli	University of Parma, Italy
Andrea Lodi	University of Bologna, Italy
Manuel López-Ibáñez	IRIDIA, Université Libre de Bruxelles, Belgium
Vittorio Maniezzo	University of Bologna, Italy
Yannis Marinakis	Technical University of Crete, Greece
Silvano Martello	University of Bologna, Italy
Roshan Joy Martis	St. Joseph Engineering College, India
Antonio D. Masegosa	University of Deusto, Spain
Daniel Merkle	University of Southern Denmark, Denmark
Martin Middendorf	University of Leipzig, Germany
Amir Nakib	Université Paris Est, France
Ferrante Neri	De Montfort University, UK
Frank Neumann	University of Adelaide, Australia
Eneko Osaba	University of Deusto, Spain
Djamila Ouelhadj	University of Portsmouth, UK
Luis Paquete	University of Coimbra, Portugal
Panos M. Pardalos	University of Florida, USA
Andrew Parkes	University of Nottingham, UK
Mario Pavone	University of Catania, Italy
David Pelta	University of Granada, Spain
Camelia Pintea	Technical University Cluj-Napoca, Romania
Günther Raidl	Vienna University of Technology, Austria
Helena Ramalhinho Lourenço	University Pompeu Fabra, Spain
Andreas Reinholz	German Aerospace Center, Germany
Mauricio G.C. Resende	MOP, Amazon.com, USA
Celso C. Ribeiro	Universidade Federal Fluminense, Brazil
Andrea Roli	University of Bologna, Italy
Andrea Schaerf	University of Udine, Italy
Marc Sevaux	Université de Bretagne-Sud, Lorient, France
Patrick Siarry	Université Paris-Est Créteil Val-de-Marne, France
Roman Słowiński	Poznań University of Technology, Poland
Kenneth Sörensen	University of Antwerp, Belgium
Maria Grazia Speranza	University of Brescia, Italy
Thomas Stützle	IRIDIA, Université Libre de Bruxelles, Belgium
Éric Taillard	University of Applied Sciences of Western Switzerland, Switzerland
Guy Theraulaz	CRCA, CNRS, Université Paul Sabatier, France
Paolo Toth	University of Bologna, Italy
José Luis Verdegay	University of Granada, Spain
Luis Nunes Vicente	University of Coimbra, Portugal
Daniele Vigo	University of Bologna, Italy
Stefan Voss	University of Hamburg, Germany
Alan Winfield	University of the West England, UK

External Reviewers

Olinto Araújo	Universidade Federal de Santa Maria, Brazil
Alan Freitas	Federal University of Ouro Preto, Brazil
Puca Huachi Penna	Universidade Federal Fluminense, Brazil
Andrea Raiconi	University of Salerno, Italy

Contents

Finding Uniquely Hamiltonian Graphs
of Minimum Degree Three
with Small Crossing Numbers

Benedikt Klocker$^{(\boxtimes)}$, Herbert Fleischner, and Günther R. Raidl

Institute of Computer Graphics and Algorithms, TU Wien,
Favoritenstraße 9–11/1861, 1040 Vienna, Austria
{klocker,fleischner,raidl}@ac.tuwien.ac.at

Abstract. In graph theory, a prominent conjecture of Bondy and
Jackson states that every uniquely hamiltonian planar graph must have
a vertex of degree two. In this work we try to find uniquely hamiltonian
graphs with minimum degree three and a small crossing number by minimizing the number of crossings in an embedding and the number of
degree-two vertices. We formalize an optimization problem for this purpose and propose a general variable neighborhood search (GVNS) for
solving it heuristically. The several different types of used neighborhoods also include an exponentially large neighborhood that is effectively searched by means of branch and bound. To check feasibility of
neighbors we need to solve hamiltonian cycle problems, which is done
in a delayed manner to minimize the computation effort. We compare
three different configurations of the GVNS. Although our implementation could not find a uniquely hamiltonian planar graph with minimum
degree three disproving Bondy and Jackson's conjecture, we were able to
find uniquely hamiltonian graphs of minimum degree three with crossing
number four for all number of vertices from 10 to 100.

Keywords: Variable neighborhood search · Uniquely Hamiltonian
graphs · Combinatorial optimization

1 Introduction

A lot of research in graph theory focuses on hamiltonian cycles. The problem
of finding hamiltonian cycles is well studied in theoretical aspects [12] and in
practical aspects as a special case of the traveling salesman problem [2]. An
interesting topic in graph theory is the question of how many hamiltonian cycles
a given graph has. A simpler version of this question only asks if there is exactly
one hamiltonian cycle in a given graph.

In this paper we will only be concerned with undirected simple graphs and
just write graph for this type of graphs. A graph $G = (V, E)$ is uniquely hamiltonian if and only if it contains exactly one hamiltonian cycle, i.e., a cycle visiting

This work is supported by the Austrian Science Fund (FWF) under grant P27615.

© Springer International Publishing Switzerland 2016
M.J. Blesa et al. (Eds.): HM 2016, LNCS 9668, pp. 1–16, 2016.
DOI: 10.1007/978-3-319-39636-1_1

each node exactly once. A graph G is planar if and only if it has a planar embedding, i.e., it can be drawn in the Euclidean plane without any crossing edges. The crossing number $cr(G)$ of a graph G is the smallest possible number of crossings in an embedding of G into the plane. Last but not least, a graph G has minimum degree $\delta(G) = k$ if each node has at least k incident edges.

One type of problem in the area of uniquely hamiltonian graphs is to determine for a given class of graphs if it contains a uniquely hamiltonian graph. Bondy and Jackson [4] showed that every uniquely hamiltonian graph with n vertices has a vertex with degree at most $c \log(8n) + 3$ for a small constant c. This limits the minimum degree of a uniquely hamiltonian graph (note that a better lower bound for $\delta(G)$ has been established in [1]). Bondy and Jackson proved in the same paper also a simpler statement: Every planar uniquely hamiltonian graph contains at least two vertices of degree two or three. Furthermore, they stated an interesting still unsolved conjecture that every planar uniquely hamiltonian graph contains a vertex of degree two.

In the case of non-planar graphs a question by Sheehan [21] asks whether or not a uniquely hamiltonian 4-regular graph exists. Note that a graph which only contains vertices with odd degree cannot be uniquely hamiltonian [22]. Fleischner [9] showed that in the case of multigraphs there exist 4-regular uniquely hamiltonian graphs. In more recent work Fleischner [10] constructed an infinite family of uniquely hamiltonian (simple) graphs with minimum degree four. This surprising result leads to the guess that there may also be a uniquely hamiltonian planar graph with minimum degree three which would disprove the conjecture of Bondy and Jackson. Entringer and Swart [8] constructed already in 1980 uniquely hamiltonian graphs with minimal degree three, but they are not planar.

In this paper we transform the problem of finding a uniquely hamiltonian planar graph with minimum degree three into a bi-objective optimization problem which minimizes the crossing number and the number of vertices with degree two. To solve this problem nearly optimal we propose a general variable neighborhood search (GVNS) heuristic [14]. The GVNS framework was already successfully applied to other graph theoretical problems, see, e.g., [5].

As the search space increases exponentially with increasing number of vertices a heuristic could be beneficial compared to an exact approach. In fact, before we implemented the general variable neighborhood search, we tried to apply an exact approach, which enumerates graphs of a given vertex degree in a clever way and tests if they are uniquely hamiltonian [19]. With that approach we were able to check that no uniquely hamiltonian planar graph with minimum degree three and with 22 or less vertices exists. Unfortunately, for graphs with more than 22 vertices the running time explodes because of the vast number of graphs to check.

One of our proposed neighborhood structures is exponentially large and we use a branch-and-bound procedure to find the best neighbor in its neighborhoods. This embeds the idea of a large neighborhood search [20] in our GVNS. We will see that the bottleneck in our algorithm, which consumes most of the running time, are the expensive unique hamiltonicity checks. To reduce the number of such checks we keep infeasible, not uniquely hamiltonian, solutions formally in our neighborhood. Only after finding the best neighbor we apply a Lin-Kerninghan heuristic [16] and in a later step we use Concorde [7] to check for feasibility.

In the next section we describe some transformations of the problem and formally state the resulting optimization problem. In Sect. 3 we describe our GVNS framework and the neighborhood structures in detail. The comparison of three different configurations and the experimental results are presented in Sect. 4. Finally, we conclude with Sect. 5 and propose some further work ideas.

2 Problem Description

In graph theory an important and challenging open question is whether or not a *uniquely hamiltonian planar graph with minimum degree three* (UHPG3) exists [4]. Bondy and Jackson [4] conjectured that every planar uniquely hamiltonian graph contains a vertex of degree two, i.e., that no UHPG3 exists. So far, however, neither could a UHPG3 be found nor could it be proven that none exists.

To disprove Bondy and Jackson's conjecture, it would be enough to find a planar graph with minimum degree three that contains an arbitrarily selected fixed edge $e \in E$ and exactly one hamiltonian cycle containing this edge e. This means the graph may contain also other hamiltonian cycles which do not contain the edge e. From such a graph we can remove the edge $e = (v, w)$, duplicate the remaining graph and connect the two copies by adding edges (v, v') and (w, w'), where v' and w' are the duplicates of nodes v and w, respectively. The resulting graph is then uniquely hamiltonian and still planar with minimum degree three. We call a uniquely hamiltonian graph whose Hamiltonian cycle contains the known edge e *fixed edge uniquely hamiltonian graph* (FEUHG).

In this work we concentrate on the optimization problem variant of finding a graph G with a given number of nodes $n = |V|$ that *as far as possible* corresponds to a UHPG3. To this end, we relax the conditions that the graph must be planar and must have minimum degree three and instead minimize the deviations from these properties.

To our knowledge, the problem of finding graphs that as far as possible correspond to a UHPG3 has so far not been considered in a more systematical, and in particular computational way.

More specifically, we consider the bi-objective optimization problem to find a FEUHG G together with an embedding into the plane and minimize

- the number of edge crossings and
- the number of vertices with degree two.

To obtain a single-objective optimization problem we linearly combine these two objectives with corresponding weights α and β.

Since the problem of computing the crossing number of a graph is NP-hard (see [11]), it makes sense to approximate this value. We do this by allowing only crossings between edges which are not part of the uniquely hamiltonian cycle

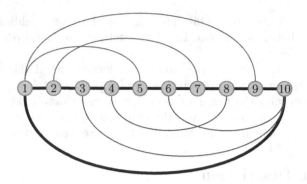

Fig. 1. Example solution with two crossings and no nodes of degree two. Edges of the hamiltonian cycles are bold and chords are colored red and blue. (Color figure online)

of the graph. This relaxation allows us to fix a hamiltonian cycle in advance and then only concentrate on the problem of adding additional edges (*chords*) between vertices that are no neighbors on the cycle. Since the added chords are not allowed to cross with an edge of the hamiltonian cycle, there are only two possibilities how to draw a chord: either outside the cycle or inside the cycle. We encode this two states, inside or outside, of a chord by two colors.

Let us assume without loss of generality that $V = \{1, \ldots, n\}$ and our predefined cycle C visits the nodes in the natural order from 1 to n before getting back to 1. If we fix the chords and their colors it is easy to derive the minimal number of crossings. To see this, we draw the nodes from 1 to n on one line, such that the hamiltonian cycle consists of this line together with a half cycle connecting the vertices n and 1. We draw now every chord as a half cycle, either above the line or below the line depending on the color of the chord. Figure 1 illustrates this construction with an example. For this construction we get that two chords (i,j) and (k,ℓ) with $1 \leq i < j \leq n$ and $1 \leq k < \ell \leq n$ cross if and only if their corresponding half cycles cross. This is the case if and only if the two chords have the same color and

$$i < k < j < \ell \text{ or } k < i < \ell < j \tag{1}$$

holds. Notice that in any drawing of a graph, two chords, which are on the same side of the graph and satisfy (1), cross at least once. This implies the optimality of our construction.

All together we get the following optimization problem. Given the cycle C over nodes $V = \{1, \ldots, n\}$, let

$$\mathcal{H} = \{(i,j) \mid i = 1, \ldots, n-2, \ j = i+2, \ldots, n\} \setminus \{(1,n)\}$$

be the set of all possible chords. A candidate solution is represented by (H, c), where $H \subseteq \mathcal{H}$ is the subset of selected chords and $c : H \rightarrow \{0,1\}$ specifies their coloring by assigning either 0 or 1.

$$\underset{\substack{H \subseteq \mathcal{H} \\ c:H \to \{0,1\}}}{\text{minimize}} \quad \alpha \cdot \sum_{(i,j) \in H} |\{(k,\ell) \in H : c(i,j) = c(k,\ell), \ i < k < j < \ell\}|$$

$$+ \ \beta \cdot \left(n - \left| \bigcup_{(i,j) \in H} \{i,j\} \right| \right) \tag{2}$$

subject to: there is no hamiltonian cycle in $C \cup H$ containing

edge $(1,2)$ and at least one chord. $\qquad$ (3)

In this formulation, there is always the same edge, namely the edge $(1,2)$, fixed. This eliminates some symmetries. Be aware that a vertex has degree three or more if and only if it is incident to at least one used chord. Therefore, the union in the second part of the objective expression (2) contains exactly all vertices of degree three or more. In the following we refer to constraint (3) also simply as *unique hamiltonicity*.

The problem of checking if a given (planar) graph contains a hamiltonian cycle is NP-complete. Therefore, the problem of checking if a given (planar) graph contains no hamiltonian cycle is in coNP. If coNP $\neq$ NP this would also imply that the latter problem is not in NP. Thus, checking only feasibility in our model is already a hard problem.

3 General Variable Neighborhood Search with Delayed Feasibility Checking

In this section we will present our algorithmic approach to tackle the optimization problem described in Sect. 2. As mentioned checking only feasibility of an instance in our model is already a hard problem. Therefore, the use of heuristics, instead of an exact algorithm, appears appropriate. We propose a *general variable neighborhood search* (GVNS) as framework to solve the problem heuristically [13,14] and combine it with Concorde [7] for checking feasibility w.r.t. unique hamiltonicity. Remember that a solution is represented by the set of chords H and the associated coloring c for each chord in H. We start with the empty initial solution $H = \emptyset$, i.e., just the cycle C without any chords, as this is always a feasible solution.

The GVNS contains a *variable neighborhood descent* (VND) for locally improving candidate solutions in systematic ways according to five different types of neighborhood structures, and a parameterized shaking neighborhood structure for diversifying the search. In the following we describe these neighborhood structures and the corresponding search algorithms.

3.1 VND Neighborhoods

The following types of neighborhoods and respective algorithms to search them are considered within the VND. Different specific configurations will be considered, which will be described in Sect. 3.2. In general, if an improved solution is achieved within one neighborhood, the VND restarts with its first neighborhood structure; otherwise it continues with the next as long as one is available. When the VND terminates, a solution is obtained that is locally optimal w.r.t. all used neighborhood structures. All these neighborhoods are searched in a best-improvement fashion, and ties are broken randomly.

Changing Color of k Chords [cchol(k)]. This neighborhood consists of all solutions where the colors of k chords are flipped for some parameter $k \geq 1$. The size of the neighborhood is therefore m^k where m is the number of chords in the current solution. We apply this neighborhood structure for $k = 1$ and $k = 2$ since the neighborhood is relatively small and easy to search. Since only the colors of chords are changed, the structure of the solution graph stays the same, and therefore the unique hamiltonicity is still valid for each neighbor of a feasible incumbent solution. To calculate the objective gain of a neighbor incrementally we simply count the number of crossings with the old color and with the new color and take the difference.

Removing k Chords and Adding ℓ New Chords [remadd(k, ℓ)]. This neighborhood consists of all solutions where exactly k chords are removed from the current solution and ℓ new chords are added. The size of the neighborhood is therefore $m^k (M - m)^\ell$ where m is the number of chords in the current solution and M is the number of possible chords ($M = (n-2)(n-1)/2 - 1$). For $\ell > 0$ the neighborhood may contain solutions which are infeasible as they are not uniquely hamiltonian. Instead of checking feasibility immediately for each neighbor, which can be very time-expensive, we first evaluate all neighboring solutions according to our objective function, filter out any solutions that are not better than our incumbent, and sort all better solutions according to their objective function gain. Only then we consider the these solutions according to decreasing gain, check the feasibility of each w.r.t. unique hamiltonicity, and immediately return with the first, and thus best, feasible solution. To calculate the objective gain of a neighbor incrementally we count the crossings of the removed edges and the crossings of the added edges and take the difference. Additionally we have to check all vertices incident to removed and added edges if their degree decreased to two or increased to three or higher. In case there are multiple, i.e., equally good, best neighbors, all of these are checked for feasibility and one is chosen at random. This random tie breaking turned out to be crucial for the performance. The procedure of checking unique hamiltonicity will be described in Sect. 3.4. In case of $\ell = 0$ we do not need this check as a solution obtained from a feasible solution by just removing chords cannot become infeasible. The values used for k and ℓ will be described later.

Computing Optimal Crossings [compcross]. This is an exponentially large neighborhood consisting of all solutions that have the same underlying graph as the current solution but different colors on the chords. The size of this neighborhood is thus $2^m - 1$, where m is the number of chords in the current solution. Instead of a naive enumeration we search this neighborhood in an efficient way by a branch-and-bound procedure to obtain a best possible coloring of the chords. The crossings of the current solution can be used as a good initial upper bound, which is in many cases already tight. In every level of the search tree we assign one color to one chord. The sequence of chords is predefined randomly and used to determine which chord is colored next. As a branching strategy we use depth-first search. To compute a local lower bound we use the crossings of the currently assigned chords and add for every not assigned chord the minimum of crossings to the assigned chords for the two colors. As already mentioned finding the optimal crossings is an NP-hard problem, but with the described branch-and-bound procedure it can be computed relatively fast compared to the effort which is needed to check for unique hamiltonicity.

Split Crossings [splitcross]. This neighborhood is a special subset of the neighborhood remadd$(2, 2)$ which tries to resolve a crossing by splitting. More precisely for every crossing pair of chords (i, j) and (k, ℓ) with $i < k < j < \ell$ we construct a new solution by removing these two edges and adding the chords (i, ℓ) and (k, j) instead. There are two exceptions: if $j = k + 1$ the edge (k, j) would be no chord and therefore this case is skipped. Similarly, case $i = 1$ and $\ell = n$ is excluded since these two vertices are already connected in the original hamiltonian cycle. If the chords (i, ℓ) or (k, j) already exist in the current solution, this neighbor is skipped. If the newly added edges do not generate other crossings and the new graph is still uniquely hamiltonian we get a solution with one crossing less. The size of this neighborhood is equal to the number of crossings in the current solution and is therefore typically a small subset of remadd$(2, 2)$.

Merge Crossings [mergecross]. This neighborhood is a special subset of the neighborhood remadd$(1, 2)$ which tries to resolve a crossing by merging the crossing point into one of the neighboring vertices. We generate up to four new solutions for every pair of crossing chords (i, j) and (k, ℓ) with $i < k < j < \ell$ by applying the following operations:

1. Remove (i, j) and add (i, k) and (k, j).
2. Remove (i, j) and add (i, ℓ) and (j, ℓ).
3. Remove (k, ℓ) and add (i, k) and (i, ℓ).
4. Remove (k, ℓ) and add (k, j) and (j, ℓ).

Cases where the edges to be added do not correspond to valid chords or where they already exist in the solution are skipped again.

3.2   VND Neighborhood Selection

In our experiments in Sect. 4, we will consider three different configurations of VND neighborhood, which are shown in Table 1. The VND considers the stated specific neighborhoods in the listed order.

Table 1. Three different neighborhood structure sets used to compare with each other

Slim set	Medium set	Thick set
1. cchol(1)	1. cchol(1)	1. cchol(1)
2. cchol(2)	2. cchol(2)	2. cchol(2)
3. remadd(1, 0)	3. remadd(1, 0)	3. remadd(1, 0)
4. remadd(0, 1)	4. remadd(0, 1)	4. remadd(0, 1)
	5. splitcross	5. splitcross
	6. mergecross	6. mergecross
	7. compcross	7. compcross
	8. remadd(1, 1)	8. remadd(1, 1)
	9. remadd(2, 1)	9. remadd(2, 1)
		10. remadd(2, 2)

It is important to notice that all neighborhoods choose the candidates randomly if their improvements are the same. This implies that the slim neighborhood set is still capable of reaching all feasible solutions. The neighborhoods which do not require rechecking unique hamiltonicity are listed before the expensive neighborhoods which require rechecking. The only exception of this rule is the compcross neighborhood which may also need significant time for larger graphs as it solves an NP-problem by branch-and-bound.

3.3 Shaking

To diversify the search, the GVNS applies the following shaking operation parameterized by $k \in \{3, \ldots, \lfloor \frac{n}{2} \rfloor\}$. Considering the chords in an order of non-increasing number of crossings, k chords are deleted from the current solution. If there are less than k chords in the current solution we delete them all. Be aware that there are in general multiple chords with the same number of crossings (most of them will have no crossings in a good solution), and they are then considered in a random order. Thus, there also is a significant randomization involved. Since we do not know in advance how many chords a solution will have, it is so far not guaranteed that in principle our GVNS can reach every possible solution from an incumbent solution. Therefore, we add as last shaking operation the removal of all chords. In other words this last shaking operation corresponds to a complete restart of the GVNS, and we trust on the VND to add chords again.

Thus, our shaking neighborhoods do not contain the complete solution space but rather a cone of all solutions we can get by removing chords from the current solution. Just removing chords in the shaking has the advantage that we will never violate unique hamiltonicity, and thus we always get a feasible solution efficiently. It would be difficult to generate a random solution in the complete feasible solution space since we would have to check for unique hamiltonicity until we find a solution satisfying it. This would cost a lot of time which is not our intention behind shaking. However, our overall approach guarantees that every feasible solution can in principle be found by descending from one of the solutions generated by shaking. This can easily be seen by the fact that every solution can be constructed by adding chords from the empty solution.

3.4 Checking Unique Hamiltonicity

In this section we describe the procedure we apply to check if a given solution is uniquely hamiltonian, i.e., satisfies (3). The running time of this procedure is crucial for the success of the algorithm since it is the bottleneck of the GVNS as we will see in the experimental results in Sect. 4.

As we only check unique hamiltonicity for neighbors which would improve the current solution we can descend to the neighbor whenever the condition is satisfied. This means that the number of procedure calls where the condition is satisfied corresponds to the number of local improvements. However, it is possible that any number of neighbors get checked before one is found which satisfies condition (3). Therefore, we do not have a better bound on the number of negative procedure calls than the current neighborhood size.

Since the hamiltonian cycle problem is a special case of the well studied travelling salesman problem, there already exist a lot of practically efficient algorithms to approach the hamiltonian cycle problem. To model the hamiltonian cycle problem as a traveling salesman problem one simply assigns all pairs of nodes corresponding to edges in the graph, i.e., E, unit costs and all other pairs of nodes larger costs. The question whether or not a hamiltonian cycle exists is then equivalent to the question whether or not a tour with costs $|V|$ exists. If we want to fix a subset of edges $E' \subseteq E$, then we can give them zero costs. The question if a hamiltonian cycle containing all edges in E' exists is then equivalent to the question if a tour with costs $|V| - |E'|$ exists. Thus, we can also model a fixed edge hamiltonian cycle problem.

To check condition (3), more specifically we need to check if a hamiltonian cycle containing edge $(1, 2)$ and edge e for any chord $e \in H$ exists. This means, we have to solve $|H|$ traveling salesman problems before we know for sure that condition (3) is satisfied. If at some point we find a hamiltonian cycle we can stop and know that the condition is not satisfied.

Clearly, solving $|H|$ traveling salesman problems would be very time-expensive. Fortunately, we can apply an improvement in our case that takes into account that the considered candidate solution graph is a neighbor of our current solution for which we already know that it satisfies condition (3). Remember that the only situation where we have to recheck the condition is for neighbors where we removed k chords and added $\ell > 0$ new chords. In this situation it is sufficient to check if there exists a hamiltonian cycle containing the edge $(1, 2)$ and at least one of the *newly added* chords. Therefore, we only have to solve ℓ traveling salesman problems to perform this check.

As we already mentioned there may be many more negative procedure calls than positive ones and therefore we need a solver which is able to find hamiltonian cycles fast. Thus, we decided to use a heuristic to find hamiltonian cycles. We use Helsgaun's version of the Lin-Kerninghan heuristic, which is faster than exact approaches but can still solve many problems to optimality [16]. If one run of the LKH heuristic does not find a hamiltonian cycle we apply ten further runs of the heuristic. If it still does not find a hamiltonian cycle we assume that the graph does not contain anyone. To avoid that the algorithm returns an infeasible solution as an optimal solution at the end we use Concorde [7] to check for hamiltonian cycles. As Concorde needs much more time than the Lin-Kerninghan heuristic, we only call Concorde whenever a neighbor would lead to a new best solution and the Lin-Kerninghan heuristic did not find a cycle in any run.

This means it may happen that an infeasible neighbor gets visited if it is no new best solution. In this case Lin-Kerninghan calls, where we assume that the current solution is feasible, are not correct anymore. This is no problem since as soon as the search visits a neighbor which would be a new best solution, Concorde gets applied and shows that the neighbor is infeasible. As we will see in Sect. 4 this situation will almost never happen for small vertex degrees.

Further Improvements. To further improve the running time of checking unique hamiltonicity we exploit the fact that only small parts of the graph change when doing local improvements. Obviously the same hamiltonian cycles may frequently appear in the investigated graphs having only small differences. The idea is now to store found hamiltonian cycles in an appropriate data structure which allows us to check for a new graph if it contains a cycle from the data structure efficiently. If searching through this data structure can be done reasonably fast, this will improve the overall runtime. We can represent a cycle or a whole graph by the set of its edges. Thus, we need a data structure which stores sets and can compute subset queries quickly. This problem is known as the containment query problem [6]. It is the complementary problem of the better known subset query problem, which furthermore corresponds to the well known partial match problem [18].

For our purposes we used a trie data structure presented in [3]. Note that we only store the set of chords used in the hamiltonian cycle. Checking if a subset exists in such a data structure needs exponential time in the worst case. Nevertheless, it is still much faster to search in this data structure than searching a hamiltonian cycle in practice, as we will see in Sect. 4. There would also be more sophisticated data structures for storing sets (see for example [15]). In our case we do not need such complex data structures since our practical tests indicated that the simple data structure's time consumption is almost neglectable in comparison to the effort for finding hamiltonian cycles in the remaining cases.

4   Computational Results

Our VNS-approach is implemented in C++ and compiled with g++ 4.8.4. We used the LKH-heuristic implementation provided in [17]. The Concorde implementation from [7] uses CPLEX 12.6.2 for solving. All tests were performed on a single core of an Intel Xeon E5540 processor with 2.53 GHz and 10 GB RAM. The input of our algorithm is simply the number of vertices $n \in \mathbb{N}$. For all tests we used the weights $\alpha = 0.25$ and $\beta = 1$ (see (2)). This implies that all solution graphs with an objective value smaller than 1 have minimum degree three.

As the instances we used different n values between 10 and 100. We ran all three configurations (see Table 1) for every instance 20 times with different seed values and a maximal execution time of 3600 seconds per run. In Table 2 we see the results for the three different configurations for different instances n. The columns *best* contain the best value found in all 20 runs for one configuration. The columns *avg.* contain the averages of the results over the 20 runs and the columns $t[s]$ *med.* contain the medians of the times until the best solution was found in each run in seconds. For every instance and every of the three column types we marked the best value of the three configurations by displaying it bold.

Table 2. Results for the three different configurations

	Slim set			Medium set			Thick set		
n	*best*	*avg.*	$t[s]$ *med.*	*best*	*avg.*	$t[s]$ *med.*	*best*	*avg.*	$t[s]$ *med.*
10	**0.5**	**0.5**	56.83	**0.5**	**0.5**	37.77	**0.5**	**0.5**	**21.33**
15	**0.5**	**0.5**	5.45	**0.5**	**0.5**	**4.92**	**0.5**	**0.5**	15.77
20	**0.5**	**0.5**	13.23	**0.5**	**0.5**	**12.29**	**0.5**	**0.5**	21.71
25	**0.5**	**0.5**	50.25	**0.5**	**0.5**	48.92	**0.5**	**0.5**	82.42
30	**0.5**	**0.5**	234.6	**0.5**	**0.5**	**75.03**	**0.5**	**0.5**	301.92
35	**0.5**	**0.5**	**139.81**	**0.5**	**0.5**	209.68	**0.5**	**0.5**	442.44
40	**0.5**	**0.5**	369.86	**0.5**	**0.5**	**277.22**	**0.5**	0.6	126.32
45	**0.5**	0.55	1,020.1	**0.5**	**0.53**	714.29	**0.5**	0.7	1,510.86
50	**0.5**	0.74	144.2	**0.5**	**0.56**	1,007.63	**0.5**	0.81	321.41
60	**0.5**	0.93	19.52	**0.5**	**0.83**	43.22	**0.5**	0.94	99.78
70	**0.5**	0.98	35.76	**0.5**	**0.76**	238.53	**0.5**	1	306.56
80	**0.5**	0.96	71.46	**0.5**	**0.9**	68.08	1	1.2	564.71
90	0.75	1.04	128.52	**0.5**	**0.95**	130.56	1	1.46	1,260.15
100	1	1	183.33	**0.5**	**0.96**	173.95	**0.5**	1.66	570.27

As we can see, on average, the medium set found better solutions than the other two configurations. Note that the running times until the best solution was found are only comparable if the corresponding solutions are equally good. Therefore, it makes no sense to compare the time medians for the instances with $n \geq 45$. To verify that the medium solution performs better than the other two solutions we applied a Wilcoxon signed-rank test. We use a p-value of 5 % for the significance value. As a result we get that the medium set computed for the instances $n = 50, 60, 70$ significantly better results than the slim set and for all instances with $n \geq 40$ significantly better results than the thick set. For all other instances the difference was not significant. We also compared the slim set with the thick set and interestingly the slim set computed for the instances $n = 40, 45, 80, 90, 100$ significantly better solutions than the thick set.

To compare the running times until the best solution was found we also applied the test for the running times, but only for the instances with $n \leq 40$. If we compare the medium set and the slim set there is only for the instance $n = 30$ a significant difference, where the medium set is significantly faster than the slim set. If we compare the slim or the medium set with the thick set, we get that both are for the instances with $15 \leq n \leq 35$ significantly faster than the thick set. For $n = 10$ the thick set is the fastest on average and compared to the slim set it is also significantly faster.

From these tests we get the intuition that the neighborhood remadd(2,2) is too large and not beneficial for graphs of medium size or larger sizes. Only for graphs with size around $n = 10$ the neighborhood is small enough to be beneficial. We need, however, more neighborhoods than only adding and removing chords, as in the slim set, to find good solutions for larger instances. It is also interesting that the slim set has no significant speed gain compared to the medium set. From the *best* column of the medium set we see that we found a solution with an objective of 0.5 for all given instances. This means that these solution graphs have minimum degree three and exactly two crossings in the planar embedding. We also applied some test runs for all other n values between 10 and 100 and found a solution with an objective of 0.5 for every $n \in \{10, \ldots, 100\}$.

Note that the solutions of our problem, as we stated it, are not uniquely hamiltonian, they are FEUHG. That means to get a uniquely hamiltonian graph we need to duplicate it as described in Sect. 2, but then the crossings get also duplicated. Therefore, all our solutions with an objective of 0.5 induce uniquely hamiltonian graphs with minimum degree three which have embeddings with four crossings. Therefore, our results impose the following question:

Does there exist a uniquely hamiltonian graph with minimal degree three and a crossing number smaller than four?

Table 3 contains the number of performed local search improvements for each neighborhood and each instance with the medium set configuration. The numbers are averaged over all 20 runs with 20 different seeds. The column name $ra(x, y)$ stands for remadd(x, y), *splitcr* for splitcrossing, *mergecr* for mergecrossing and *compcr* for compcrossing.

Since the shake operations only remove chords, the neighborhood remadd(0, 1) which only adds one chord is applied after shaking several times. This explains why this neighborhood is used much more often than the other neighborhoods.

To find out which part of the algorithm consumes the most time we can identify three subroutines which have an exponential worst case running time. The first one is Concorde, which gets applied whenever the search finds a new best solution for which the LKH-heuristic did not find a second hamiltonian cycle. The second one is the solver of the containment query problem, which uses a trie data structure to check if a cycle which was already found is contained

Table 3. Number of improvements found in the different neighborhood structures for the medium set configuration

n	$cchol(1)$	$cchol(2)$	$ra(1,0)$	$ra(0,1)$	$splitcr$	$mergecr$	$compcr$	$ra(1,1)$	$ra(2,1)$
10	2,667	60	0	314,452	677	197	0	3,898	4
15	6,551	3,872	19	279,279	522	856	3,126	3,202	204
20	3,904	1,887	19	202,046	1,599	1,371	1,732	2,211	128
25	2,676	1,179	26	133,997	1,293	1,287	1,311	1,675	86
30	1,828	646	34	92,814	1,201	1,152	895	1,334	63
35	1,257	501	23	63,717	1,013	958	550	1,054	44
40	999	375	24	48,623	844	828	482	869	39
45	750	248	32	36,387	769	771	284	794	30
50	595	193	20	28,707	661	640	232	656	26
60	365	107	25	18,612	553	564	99	559	17
70	264	75	15	12,953	398	410	77	390	13
80	201	53	17	9,573	330	336	50	328	9
90	147	37	17	7,184	258	283	39	263	8
100	123	29	10	5,891	221	229	27	212	5

in the current candidate. The third one is the branch and bound procedure to calculate the optimal colors of the chords such that they have a minimal number of crossings.

Table 4 lists running time information and additional information for these three subroutines and the LKH subroutines with the medium set configuration. The column *HC-Checks* contains the number of graphs for which we had to test the unique hamiltonicity constraint and the *UHGs* column contains the number of graphs which satisfied the unique hamiltonicity constraint. The column *calls* contains the number of Concorde calls and the column *rate* contains the number of graphs which got discarded by a containment query relative to the overall number of discarded graphs. The columns $t[s]$ contain the overall time used for all corresponding calls in seconds. The time columns represent median values and the other columns represent average values over all different seeds. As we can see the three mentioned subroutines are in total extremely fast compared to the LKH subroutines which have to get applied very often. Therefore, the LKH subroutines are clearly the bottleneck of the whole VNS. Another interesting fact is that Concorde never found a hamiltonian cycle in any of the runs. This means that whenever LKH did not find a second hamiltonian cycle the graph was in fact uniquely hamiltonian.

The fact that the rates of the containment queries increase with increasing n can be explained as follows. First of all, for two solutions which are very similar it is more likely that they both contain the same hamiltonian cycle than for solutions which are completely different. This implies that as long as the search

Table 4. Hard subproblem statistics for the medium set configuration

	LKH			Concorde		Containment Queries		B&B
n	HC-Checks	UHGs	t[s]	calls	t[s]	rate	t[s]	t[s]
10	1,467,893	255,790	1,356	7	0	4.8%	3	0
15	1,607,901	258,962	2,480	10	0	5.5%	4	0
20	1,456,778	200,372	3,230	14	0	7.5%	7	0
25	1,151,036	137,859	3,421	17	0	10.1%	8	0
30	929,409	99,379	3,491	21	1	12.3%	9	0
35	761,117	70,547	3,512	24	1	14.5%	10	0
40	653,680	55,017	3,518	26	1	17.2%	11	0
45	570,683	42,607	3,521	30	1	19.3%	12	0
50	493,993	34,275	3,523	33	2	21.3%	13	0
60	425,704	23,800	3,523	40	3	28.5%	14	0
70	323,338	16,864	3,528	46	5	30.6%	14	1
80	277,038	12,984	3,525	52	9	34.2%	13	1
90	246,591	10,072	3,515	60	11	39.1%	17	2
100	204,408	8,324	3,510	65	14	41.3%	21	4

is concentrated in a local area the containment queries will be very effective. Only the shaking methods guide the search out of such a local area. The simple fact that for larger n the search in the local neighborhoods needs longer implies that it can do less shaking than for smaller n. Therefore, the containment queries are more effective for larger n.

We want to mention that we tested the algorithm also for $n > 100$. For these instances the running times of the subproblems exploded and the VNS could do only few iterations in reasonable time which lead to poor quality solutions.

5   Conclusions and Future Work

In this paper we presented a new optimization problem for finding uniquely hamiltonian graphs of minimum degree three with small crossing numbers. We proposed a general variable neighborhood search framework to solve the problem heuristically. We proposed different neighborhood structures including one large neighborhood and different configurations which we compared in experimental tests. The bottleneck of the proposed algorithm is checking unique hamiltonicity for every neighbor, to stay in the feasible area. With an implementation of this framework we were able to find uniquely hamiltonian graphs of minimum degree three with only four crossings for many different instances, which naturally gives rise to the question if we can do better.

Future work may be to develop an exact algorithm for the proposed problem and compare the two algorithms for small instances. One problem of the proposed heuristic is that the constraint of unique hamiltonicity is completely independent from the objective function. If we could measure how promising a graph is according to the unique hamiltonicity constraint we could better guide the search. Furthermore, it would be interesting to test if other heuristics than LKH or other variants of LKH for checking unique hamiltonicity perform better.

References

1. Abbasi, S., Jamshed, A.: A degree constraint for uniquely Hamiltonian graphs. Graphs and Combinatorics **22**(4), 433–442 (2006)
2. Applegate, D.L., Bixby, R.E., Chvátal, V., Cook, W.J.: The Traveling Salesman Problem: A Computational Study. Princeton University Press, Princeton (2011)
3. Bevc, S., Savnik, I.: Using tries for subset and superset queries. In: Proceedings of the ITI 2009, pp. 147–152 (2009)
4. Bondy, J.A., Jackson, B.: Vertices of small degree in uniquely Hamiltonian Graphs. J. Comb. Theory Ser. B **74**(2), 265–275 (1998)
5. Caporossi, G., Hansen, P.: Variable neighborhood search for extremal graphs: 1 The AutoGraphiX system. Discrete Math. **212**(1–2), 29–44 (2000)
6. Charikar, M., Indyk, P., Panigrahy, R.: New algorithms for subset query, partial match, orthogonal range searching, and related problems. In: Widmayer, P., Triguero, F., Morales, R., Hennessy, M., Eidenbenz, S., Conejo, R. (eds.) ICALP 2002. LNCS, vol. 2380, pp. 451–462. Springer, Heidelberg (2002)
7. Cook, W.: Concorde TSP Solver (2011). http://www.math.uwaterloo.ca/tsp/ concorde/. Accessed on 31 Jan 2016
8. Entringer, R.C., Swart, H.: Spanning cycles of nearly cubic graphs. J. Comb. Theory Ser. B **29**(3), 303–309 (1980)
9. Fleischner, H.: Uniqueness of maximal dominating cycles in 3-regular graphs and of Hamiltonian cycles in 4-regular graphs. J. Graph Theory **18**(5), 449–459 (1994)
10. Fleischner, H.: Uniquely Hamiltonian graphs of minimum degree 4. J. Graph Theory **75**(2), 167–177 (2014)
11. Garey, M., Johnson, D.: Crossing number is NP-complete. SIAM J. Algebraic Discrete Methods **4**(3), 312–316 (1983)
12. Gould, R.J.: Advances on the Hamiltonian problem-a survey. Graphs and Combinatorics **19**(1), 7–52 (2003)
13. Hansen, P., Mladenović, N.: An introduction to variable neighborhood search. In: Voss, S., et al. (eds.) Metaheuristics, Advances and Trends in Local Search Paradigms for Optimization, pp. 433–458. Kluwer, Dordrecht (1999)
14. Hansen, P., Mladenović, N.: A tutorial on variable neighborhood search. Technical report G-2003-46, GERAD, July 2003
15. Helmer, S., Aly, R., Neumann, T., Moerkotte, G.: Indexing set-valued attributes with a multi-level extendible hashing scheme. In: Wagner, R., Revell, N., Pernul, G. (eds.) DEXA 2007. LNCS, vol. 4653, pp. 98–108. Springer, Heidelberg (2007)
16. Helsgaun, K.: Effective implementation of the Lin-Kernighan traveling salesman heuristic. Eur. J. Oper. Res. **126**(1), 106–130 (2000)
17. Helsgaun, K.: LKH (2012). http://www.akira.ruc.dk/~keld/research/LKH/. Accessed 03 Feb 2016
18. Jayram, T.S., Khot, S., Kumar, R., Rabani, Y.: Cell-probe lower bounds for the partial match problem. In: Proceedings of the Thirty-Fifth Annual ACM Symposium on Theory of Computing, STOC 2003, pp. 667–672. ACM, New York, (2003)

19. Klocker, B., Raidl, G.: Finding uniquely hamiltonian graphs with minimal degree three. Technical report, Algorithms and Complexity Group, TU Wien (2016)
20. Pisinger, D., Ropke, S.: Large neighborhood search. In: Gendreau, M., Potvin, J.-Y. (eds.) Handbook of Metaheuristics, pp. 399–419. Springer US, London (2010)
21. Sheehan, J.: The multiplicity of Hamiltonian circuits in a graph. In: Recent Advances in Graph Theory, pp. 477–480 (1975)
22. Thomason, A.G.: Hamiltonian cycles and uniquely edge colourable graphs. Ann. Discrete Math. **3**, 259–268 (1978). Advances in Graph Theory

Construct, Merge, Solve and Adapt: Application to Unbalanced Minimum Common String Partition

Christian Blum[1,2]([envelope])

[1] Department of Computer Science and Artificial Intelligence,
University of the Basque Country UPV/EHU, San Sebastian, Spain
christian.c.blum@gmail.com
[2] IKERBASQUE, Basque Foundation for Science, Bilbao, Spain

Abstract. In this paper we present the application of a recently proposed, general, algorithm for combinatorial optimization to the unbalanced minimum common string partition problem. The algorithm, which is labelled CONSTRUCT, MERGE, SOLVE & ADAPT, works on sub-instances of the tackled problem instances. At each iteration, the incumbent sub-instance is modified by adding solution components found in probabilistically constructed solutions to the tackled problem instance. Moreover, the incumbent sub-instance is solved to optimality (if possible) by means of an integer linear programming solver. Finally, seemingly unuseful solution components are removed from the incumbent sub-instance based on an ageing mechanism. The results obtained for the unbalanced minimum common string partition problem indicate that the proposed algorithm outperforms a greedy approach. Moreover, they show that the algorithm is competitive with CPLEX for problem instances of small and medium size, whereas it outperforms CPLEX for larger problem instances.

1 Introduction

The drawback of exact solvers in the context of combinatorial optimization problems is often that they are not applicable to problem instances of realistic sizes. When small problem instances are considered, however, exact solvers are often extremely efficient. This is because a considerable amount of time, effort and expertise has gone into the development of exact solvers. As examples consider general-purpose integer linear programming solvers such as CPLEX and Gurobi. Having this in mind, recent research efforts focused on ways of making use of

This work was supported by project TIN2012-37930-C02-02 (Spanish Ministry for Economy and Competitiveness, FEDER funds from the European Union). Additionally, we acknowledge support from IKERBASQUE. Our experiments have been executed in the High Performance Computing environment managed by RDlab (http://rdlab.cs.upc.edu) and we would like to thank them for their support.

© Springer International Publishing Switzerland 2016
M.J. Blesa et al. (Eds.): HM 2016, LNCS 9668, pp. 17–31, 2016.
DOI: 10.1007/978-3-319-39636-1_2

exact solvers within heuristic frameworks even in the context of large problem instances. A recently proposed algorithm labelled CONSTRUCT, MERGE, SOLVE & ADAPT (CMSA) [1,3] falls into this line of research. The algorithm works as follows. At each iteration, solutions to the tackled problem instance are generated in a probabilistic way. The solution components found in these solutions are then added to an incumbent sub-instance of the original problem instance. Subsequently, an exact solver such as, for example, CPLEX is used to solve the incumbent sub-instance to optimality. Moreover, the algorithm makes use of a mechanism for deleting seemingly useless solution components from the incumbent sub-instance. This is done in order to avoid that these solution components slow down the exact solver when applied to the sub-instance.

In this work we apply the CMSA algorithm to the unbalanced minimum common string partition problem (UMCSP) [4]. This problem, which is NP-hard, is a generalization of the well-known minimum common string partition problem (MCSP) [5]. The UMCSP seems to be well-suited for being tackled with CMSA, because the integer linear programming (ILP) model that we present in this work (see Sect. 2) contains an exponential number of binary variables and can, therefore, only be solved to optimality in the context of problem instances of small and medium size. The obtained results show that, indeed, the application of CMSA obtains state-of-the-art results, especially in the context of large problem instances.

The remaining part of the paper is organized as follows. In Sect. 2 we provide a technical description of the unbalanced minimum common string partition problem. Moreover, we describe the first ILP model for this problem. Next, in Sect. 4, the application of CMSA to the tackled problem is outlined. Finally, Sect. 5 provides an extensive experimental evaluation and Sect. 6 offers a discussion and an outlook to future work.

2   Unbalanced Minimum Common String Partition

The UMCSP problem can technically be described as follows. Given is an input string s^1 of length n_1 and an input string s^2 of length n_2, both over the same finite alphabet Σ. A valid solution to the UMCSP problem is obtained by partitioning s^1 into a set P_1 of non-overlapping substrings, and s^2 into a set P_2 of non-overlapping substrings, such that exists a set S with $S \subseteq P_1$ and $S \subseteq P_2$ and no letter $a \in \Sigma$ is simultaneously present in a string $x \in P_1 \backslash S$ and a string $y \in P_2 \backslash S$. Henceforth, given P_1 and P_2 let us denote the largest subset S such that the above-mentioned condition holds by S^*. The objective function value of a solution (P_1, P_2) is then $|S^*|$. The goal consists in finding a solution (P_1, P_2) such that $|S^*|$ is minimal.

Consider the following example. Given are DNA sequences $s^1 = \mathbf{AAGACTG}$ and $s^2 = \mathbf{TACTAG}$. A trivial valid solution can be obtained by partitioning both strings into substrings of length 1, that is, $P_1 = \{\mathbf{A, A, A, C, T, G, G}\}$ and $P_2 = \{\mathbf{A, A, C, T, T, G}\}$. In this case, $S^* = \{\mathbf{A, A, C, T, G}\}$, and the objective function value is $|S^*| = 5$. However, the optimal solution, with objective function value 2, is $P_1 = \{\mathbf{ACT}, \mathbf{AG}, \mathbf{A}, \mathbf{G}\}$, $P_2 = \{\mathbf{ACT}, \mathbf{AG}, \mathbf{T}\}$ and $S^* = \{\mathbf{ACT}, \mathbf{AG}\}$.

Note that the UMCSP problem [4] is a generalization of the well-known minimum common string partition (MCSP) problem, which was introduced in [5] due to its relation to genome rearrangement. In fact, the MCSP problem is obtained in case the input strings s^1 and s^2 are related, that is, in case all letters appear the same number of times in s^1 and in s^2. The MCSP problem was shown to be NP-hard even in very restrictive cases [8]. Therefore, the more general UMCSP problem is also NP-hard. In contrast to the MCSP, the UMCSP has not been tackled yet by means of heuristics or metaheuristics. The only existing algorithm is a fixed-parameter approximation algorithm described in [4]. The more specific MCSP problem has been tackled by a greedy heuristic [9], an ant colony optimization approach [6,7], and probabilistic tree search [2]. Finally, the application of the CMSA algorithm to the MCSP problem (see [3]) is currently the state-of-the-art algorithm for this problem.

3   An ILP Model for the UMCSP Problem

In order to derive an ILP model for the UMCSP problem, we introduce in the following the *common block* concept, which allows to re-phrase the problem in a different way. A *common block* b_i concerning input strings s^1 and s^2 is denoted as a triple (t_i, k_i^1, k_i^2) where t_i is a string which can be found starting at position $1 \leq k_i^1 \leq n_1$ in string s^1 and starting at position $1 \leq k_i^2 \leq n_2$ in string s^2. Let $B = \{b_1, \ldots, b_m\}$ be the arbitrarily ordered set of all possible common blocks of s^1 and s^2. Moreover, given a string t over alphabet Σ, $n(t, a)$ denotes the number of occurrences of letter $a \in \Sigma$ in string t. Specifically, $n(s^1, a)$, respectively $n(s^2, a)$, are the number of occurrences of letter $a \in \Sigma$ in input string s^1, respectively s^2. Given the definition of B, a subset S of B corresponds to a valid solution to the UMCSP problem iff the following conditions hold:

1. $\sum_{b_i \in S} n(t_i, a) = \min\{n(s^1, a), n(s^2, a)\}$ for all $a \in \Sigma$. In other words, the sum of the occurrences of a letter $a \in \Sigma$ in the common blocks present in S must be equal to the minimum number of occurrences of letter a in s^1 and s^2.
2. For any two common blocks $b_i, b_j \in S$ it holds that their corresponding strings neither overlap in s^1 nor in s^2.

With these definitions we can state the following ILP model for the UMCSP problem, which uses for each common block $b_i \in B$ a binary variable x_i indicating its selection in the solution. In other words, if $x_i = 1$, the corresponding common block b_i is selected for the solution, and if $x_i = 0$, common block b_i is not selected.

$$\min \quad \sum_{i=1}^{m} x_i \tag{1}$$

$$\text{s.t.} \quad \sum_{i \in \{1,\ldots,m \mid k_i^1 \leq j < k_i^1 + |t_i|\}} x_i = 1 \qquad \text{for } j = 1, \ldots, n_1 \tag{2}$$

$$\sum_{i \in \{1,\dots,m \mid k_i^2 \leq j < k_i^2 + |t_i|\}} x_i = 1 \qquad\qquad \text{for } j = 1, \dots, n_2 \qquad (3)$$

$$\sum_{i=1}^{m} n(t_i, a) x_i = \min\{n(s^1, a), n(s^2, a)\} \qquad\qquad \text{for } a \in \Sigma \qquad (4)$$

$$x_i \in \{0, 1\} \qquad\qquad \text{for } i = 1, \dots, m$$

The objective function (1) minimizes the number of selected common blocks. Equation (2) ensure that the strings corresponding to the selected common blocks do not overlap with respect to s^1, and Eq. (3) ensure the same with respect to s^2. Finally, Eq. (4) ensure that the number of occurrences of each letter in the selected strings is equal to the minimum number of occurrences of this letter in s^1 and s^2.

4   Application of CMSA to the UMCSP Problem

The (CMSA) algorithm, whose pseudo-code is given in Algorithm 1, works as follows. It maintains an incumbent sub-instance B', which is a subset of the complete set B of common blocks. Moreover, each common block $b_i \in B$ has a non-negative age value denoted by $age[b_i]$. In an initialization step, the best-so-far solution S_{bsf} is set to NULL, indicating that no such solution exists yet, and the sub-instance B' is initilized to the empty set. Then, at each iteration a number of n_a solutions is probabilistically generated, see function ProbabilisticSolutionGeneration(B) in line 6 of Algorithm 1. The common blocks found in these solutions are added to B' and their age is re-initilalized to 0. Afterwards, an ILP solver—we used CPLEX—is applied to solve sub-instance B', if possible within the given CPU time limit, to optimality; see function ApplyExactSolver(B') in line 12 of Algorithm 1. If S'_{opt} is better than the current best-so-far solution S_{bsf}, solution S'_{opt} replaces the best-so-far solution (line 13). Next, sub-instance B' is adapted, based on solution S'_{opt} and on the age values of the common blocks. This is done in function Adapt$(B', S'_{\mathrm{opt}}, age_{\mathrm{max}})$ in line 14. In the following we outline the functions of the algorithm in more detail.

Function ProbabilisticSolutionGeneration(B): Henceforth we call $S \subset B$ a valid partial solution if the substrings corresponding to the common blocks in S do not overlap neither concerning s^1 nor concerning s^2. Furthermore, let set $Ext(S) \subset B \backslash S$ denote the set of common blocks that may be used in order to extend S such that the result is again a valid (partial) solution. Note that when $Ext(S) = \emptyset$, S corresponds to a complete solution. Given these definitions, a simple greedy heuristic—which is an extension of the greedy heuristic from [9] for the MCSP problem—starts with the empty partial solution $S := \emptyset$ and chooses at each step from $Ext(S)$ the common block with the longest substring. This greedy heuristic will henceforth be called GREEDY.

In function ProbabilisticSolutionGeneration(B) of line 6 of Algorithm 1 we make use of the following probabilistic version of GREEDY for generating solutions to the tackled problem instance. More specifically, the construction of a

---

Algorithm 1. CMSA for the UMCSP problem

1: **given:** set B corresponding to the tackled problem instance, values for parameters n_a and age_{max}
2: $S_{bsf} := $ NULL; $B' := \emptyset$
3: $age[b_i] := 0$ for all $b_i \in B$
4: **while** CPU time limit not reached **do**
5:    **for** $i = 1, \ldots, n_a$ **do**
6:       $S := $ ProbabilisticSolutionGeneration(B)
7:       **for all** $b_i \in S$ **and** $b_i \notin B'$ **do**
8:          $age[b_i] := 0$
9:          $B' := B' \cup \{b_i\}$
10:       **end for**
11:    **end for**
12:    $S'_{opt} := $ ApplyExactSolver(B')
13:    **if** $|S'_{opt}| < |S_{bsf}|$ **then** $S_{bsf} := S'_{opt}$
14:    Adapt$(B', S'_{opt}, age_{max})$
15: **end while**
16: **return** s_{bsf}

---

solution (see Algorithm 2) starts with the empty partial solution $S := \emptyset$. At each construction step, a solution component b_i from $Ext(S)$ is chosen and added to S. This is done until S is a complete solution, that is, until $|Ext(S)| = 0$. The choice of b_i is done as follows. First, a value $\delta \in [0, 1)$ is chosen uniformly at random. In case $\delta \leq d_{rate}$, b_i is chosen such that $|t_i| \geq |t_j|$ for all $b_j \in Ext(S)$, that is, one of the common blocks whose substring is of maximal size is chosen. Otherwise, a candidate list L containing the (at most) l_{size} longest common blocks from $Ext(S)$ is built, and b_i is chosen from L uniformly at random. In other words, the greediness of this procedure depends on the pre-determined values of d_{rate} (determinism rate) and l_{size} (candidate list size). Both are input parameters of the algorithm.

Function ApplyExactSolver(B'): In this function, CPLEX is applied to the ILP model outlined in Sect. 3 for solving sub-instance B'. This is achieved by replacing all occurrences of B in this ILP model with B', and by replacing m with $|B'|$.

Function Adapt$(B', S'_{opt}, age_{max})$: First, the age of each common block in $B' \backslash S'_{opt}$ is incremented while the age of each common block in $S'_{opt} \subseteq B'$ is re-initialized to zero. Then, those common blocks from B' whose age has reached the maximum component age (age_{max}) are deleted from B'. The motivation behind the aging mechanism is that common blocks which never appear in the solutions of B' returned by the exact solver should be removed from B' after some time, because they would otherwise slow down the exact solver on the long term. In contrast, common blocks which appear in the solutions returned by the exact solver seem to be useful and should therefore remain in B'.

Algorithm 2. Function ProbabilisticSolutionGeneration(B)

1: **given:** B, d_{rate}, l_{size}
2: $S := \emptyset$
3: **while** $|Ext(S)| > 0$ **do**
4:     choose a random number $\delta \in [0, 1]$
5:     **if** $\delta \leq d_{\text{rate}}$ **then**
6:         choose b_i such that $|t_i| \geq |t_j|$ for all $b_j \in Ext(S)$
7:         $S := S \cup \{b_i\}$
8:     **else**
9:         let $L \subseteq Ext(S)$ contain the (at most) l_{size} longest common blocks from $Ext(S)$
10:         choose b_i from L uniformly at random
11:         $S := S \cup \{b_i\}$
12:     **end if**
13: **end while**
14: **return** complete solution S

---

5   Experimental Evaluation

Three different solution methods are compared in the following. The first on is GREEDY, the simple, deterministic, greedy algorithm mentioned in Sect. 4 in the context of probabilistically generating solutions to the UMCSP problem. The second one is the CMSA algorithm, henceforth denoted by CMSA. And the third one is the application of IBM ILOG CPLEX v12.1 to the original problem instances, labelled CPLEX. The solution methods were implemented in ANSI C++ using GCC 4.7.3. Both in the context of CMSA and CPLEX, CPLEX was used in one-threaded mode. The experimental evaluation was performed on a cluster of PCs with Intel(R) Xeon(R) CPU 5670 CPUs of 12 nuclei of 2933 MHz and at least 40 Gb of RAM. Note that the fixed-parameter approximation algorithm described in [4] was not included in the comparison because, according to the authors of this work, the algorithm is only applicable to very small problem instances.

In the following we first describe the set of benchmark instances that we generated to test the considered solution methods. Then, we describe the tuning experiments that were performed in order to determine a proper setting for the parameters of CMSA. Finally, an exhaustive experimental evaluation is presented.

5.1   Problem Instances

For the comparison of the three considered solution methods we generated a set of 600 benchmark instances. In more detail, this benchmark set consists of 10 randomly generated instances for each combination of the *base-length* $n \in \{200, 400, \ldots, 1800, 2000\}$, the alphabet size $|\Sigma| \in \{4, 12\}$, and a so-called *length-difference* $ld \in \{0, 10, 20\}$. In the context of all instances, each letter of Σ has the same probability to appear at any of the positions of input strings s^1 and s^2.

Given a value for the base-length n and the length-difference ld, the length of s^1 is determined as $n + \lfloor (ld \cdot n)/100 \rfloor$ and the length of s^2 as $n - \lfloor (ld \cdot n)/100 \rfloor$. In other words, ld refers to the length difference between s^1 and s^2 (in percent) given a certain base-lenth n.

5.2 Tuning of CMSA

There are several parameters involved in CMSA for which well-working values must be found: (n_a) the number of solution constructions per iteration, (age_{max}) the maximum allowed age of common blocks, (d_{rate}) the determinism rate, (l_{size}) the candidate list size, and (t_{max}) the maximum time in seconds allowed for CPLEX per application to a sub-instance. The last parameter is necessary, because even when applied to reduced problem instances, CPLEX might still need too much computation time for solving such sub-instances to optimality. In any case, CPLEX always returns the best feasible solution found within the given computation time.

We made use of the automatic configuration tool irace [10] for the tuning of the five parameters. In fact, irace was applied to tune CMSA separately for instances of each *base-length*, which—after initial experiments—seemed to be the parameter with most influence on the algorithm performance. For each of the 10 considered base-length values, 12 tuning instances were randomly generated: two for each of the six combinations of Σ and ld. The tuning process for each alphabet size was given a budget of 1000 runs of CMSA, where each run was given a computation time limit of 3600 CPU seconds. Finally, the following parameter value ranges were chosen concerning the five parameters of CMSA:

- $n_a \in \{10, 30, 50\}$.
- $age_{max} \in \{1, 5, 10, inf\}$, where *inf* means that no common block is ever removed from sub-instance B'.

Table 1. Results of tuning CMSA with irace.

n	n_a	age_{max}	d_{rate}	l_{size}	t_{max}
200	50	10	0.0	10	480
400	50	10	0.0	10	120
600	50	10	0.0	10	240
800	50	5	0.5	10	120
1000	50	10	0.7	10	60
1200	50	5	0.5	10	120
1400	50	10	0.9	10	480
1600	50	5	0.9	10	480
1800	50	5	0.9	10	480
2000	50	10	0.9	10	480

- $d_{\text{rate}} \in \{0.0, 0.3, 0.5, 0.7, 0.9\}$, where a value of 0.0 means that the selection of the next common block to be added to the partial solution under construction is always done randomly from the candidate list, while a value of 0.9 means that solution constructions are nearly deterministic.
- $l_{\text{size}} \in \{3, 5, 10\}$.
- $t_{\text{max}} \in \{60, 120, 240, 480\}$ (in seconds).

The tuning runs with irace produced the configurations of CMSA as shown in Table 1. The most important tendencies that can be observed are the following ones. First, with growing base-length, the greediness of the solution construction grows, as indicated by the increasing value of d_{rate}. Second, the number of solution constructions per iteration is always high. Third, the time limit for CPLEX does not play any role for smaller instances. However, for larger instances the time limit of 480 seconds is consistently chosen.

5.3   Experimental Results

The numerical results are presented in Table 2 concerning all instances with $|\Sigma| = 4$, and in Table 3 concerning all instances with $|\Sigma| = 12$. Each table row presents the results averaged over 10 problem instances of the same type. For each of the three solution methods in the comparison we provide (at least) the following two columns. The first one (with heading **mean**) provides the average values of the best solutions obtained over 10 problem instances, while the second column (with heading **time**) provides the average computation time (in seconds) necessary for finding the corresponding solutions. In the case of CPLEX, this column provides two values in the form X/Y, where X corresponds to the (average) time at which CPLEX was able to find the first valid solution, and Y to the (average) time at which CPLEX found the best solution within 3600 CPU seconds. An additional column with heading **gap** provides—in the case of CPLEX—the average optimality gaps (in percent), that is, the average gaps between the upper bounds and the values of the best solutions when stopping a run. A third additional column in the case of CMSA (with heading **size** (%)) provides the average size of the sub-instances considered in CMSA in percent of the original problem instance sizes, that is, the sizes of the complete sets B of common blocks. Finally, note that the best result for each table row is marked by a gray background and the last row of each table provides averages over the whole table. Moreover, the numerical results are presented graphically in Figs. 1 and 2 in terms of the improvement of CMSA over CPLEX and GREEDY (in percent).

The results allow to make the following observations:

- Concerning the application of CPLEX to the original problem instances, the alphabet size has a strong influence on the problem difficulty. For instances with $|\Sigma| = 4$, CPLEX is only able to provide feasible solutions within 3600 CPU seconds for input strings of lengths up to 800. When $|\Sigma| = 12$, CPLEX provides feasible solutions for input strings of lengths up to 1600 (for values of $ld \in \{0, 10\}$). When $ld = 20$ CPLEX is even able to provide feasible solutions for all problem instances.

Table 2. Results for the instances with $|\Sigma| = 4$.

(a) Results for instances with $ld = 0$.

n	GREEDY		CPLEX			CMSA		
	mean	time	mean	time	gap	mean	time	size (%)
200	67.0	< 1.0	55.3	4/21	0.0	55.3	90.1	30.7
400	119.4	< 1.0	98.7	118/1445	2.1	99.4	1878.3	14.7
600	172.8	< 1.0	146.0	556/1865	6.7	145.7	2317.5	9.3
800	222.5	< 1.0	189.1	2136/3525	8.1	190.8	1837.3	5.0
1000	271.7	1.6	n.a	n.a.	n.a.	235.1	1320.3	6.1
1200	314.3	2.0	n.a	n.a.	n.a.	274.1	1837.9	3.5
1400	368.5	3.7	n.a	n.a.	n.a.	320.4	2455.8	2.6
1600	413.2	4.9	n.a	n.a.	n.a.	358.3	2875.8	2.0
1800	450.5	6.7	n.a	n.a.	n.a.	401.6	2802.1	1.7
2000	504.8	9.3	n.a	n.a.	n.a.	453.4	2166.7	1.5
avg.								

(b) Results for instances with $ld = 10$.

n	GREEDY		CPLEX			CMSA		
	mean	time	mean	time	gap	mean	time	size (%)
200	61.0	< 1.0	52.1	3/5	0.0	52.1	88.0	29.0
400	108.4	< 1.0	90.3	102/675	0.0	91.2	801.0	13.0
600	151.4	< 1.0	126.3	548/3018	2.5	127.3	1500.4	7.8
800	192.6	< 1.0	164.2	2038/3583	4.6	164.6	1513.2	3.6
1000	232.8	1.5	n.a.	n.a.	n.a.	198.3	2504.4	3.5
1200	260.8	1.9	n.a.	n.a.	n.a.	231.7	2334.2	2.2
1400	313.1	3.5	n.a.	n.a.	n.a.	267.6	2170.0	1.8
1600	346.6	4.6	n.a.	n.a.	n.a.	301.3	3114.6	1.3
1800	383.3	6.4	n.a.	n.a.	n.a.	330.9	2652.8	1.1
2000	423.1	8.8	n.a.	n.a.	n.a.	364.7	2512.3	1.0
avg.								

(c) Results for instances with $ld = 20$.

n	GREEDY		CPLEX			CMSA		
	mean	time	mean	time	gap	mean	time	size (%)
200	51.8	< 1.0	44.8	3/3	0.0	44.8	241.9	23.8
400	89.4	< 1.0	77.4	86/90	0.0	77.6	251.6	9.9
600	127.9	< 1.0	108.5	467/634	0.0	109.3	716.8	5.7
800	159.5	< 1.0	135.8	1941/2583	0.2	138.0	1009.4	2.8
1000	197.9	1.4	n.a.	n.a.	n.a.	169.6	1220.1	2.7
1200	229.9	1.7	n.a.	n.a.	n.a.	198.9	1659.6	1.7
1400	262.1	3.1	n.a.	n.a.	n.a.	229.2	2052.5	1.7
1600	294.4	4.1	n.a.	n.a.	n.a.	255.4	1830.2	1.2
1800	327.6	5.9	n.a.	n.a.	n.a.	284.5	2620.1	1.0
2000	359.0	8.1	n.a.	n.a.	n.a.	313.0	2160.0	1.0
avg.								

Table 3. Results for the instances with $|\Sigma| = 12$.

(a) Results for instances with $ld = 0$.

n	GREEDY		CPLEX			CMSA		
	mean	time	mean	time gap		mean	time	size (%)
200	107.5	< 1.0	98.1	0/0	0.0	98.1	0.3	52.0
400	204.7	< 1.0	181.2	4/7	0.0	181.8	791.5	34.0
600	290.2	< 1.0	251.8	28/599	0.0	254.9	1710.8	23.2
800	379.5	< 1.0	328.1	134/1338	1.1	331.6	1020.8	18.3
1000	467.3	< 1.0	399.5	305/2413	1.6	403.7	1682.6	18.3
1200	537.5	1.0	468.7	672/2800	2.9	469.3	2057.3	12.7
1400	624.7	2.1	543.2	1089/2641	3.6	539.7	2108.7	10.3
1600	706.0	2.6	672.2	2193/2661	12.0	610.9	1875.5	8.7
1800	794.2	3.6	n.a.	n.a.	n.a.	694.2	2300.9	8.0
2000	876.3	5.1	n.a.	n.a.	n.a.	758.1	2356.0	7.4
avg.								

(b) Results for instances with $ld = 10$.

n	GREEDY		CPLEX			CMSA		
	mean	time	mean	time gap		mean	time	size (%)
200	104.2	< 1.0	95.9	0/0	0.0	95.9	0.9	54.2
400	196.5	< 1.0	173.5	5/7	0.0	173.8	733.1	32.7
600	274.2	< 1.0	240.1	34/119	0.0	242.0	1275.3	22.7
800	354.1	< 1.0	304.9	109/903	0.2	308.3	605.0	16.0
1000	427.3	< 1.0	369.0	303/2035	0.3	373.9	1332.0	15.2
1200	502.6	1.0	433.3	550/2457	0.7	439.0	1615.6	10.8
1400	570.1	1.7	494.4	1011/2249	1.3	498.6	1146.7	9.4
1600	646.8	2.4	561.4	1738/3071	2.1	561.8	1493.3	7.1
1800	712.7	3.6	n.a.	n.a.	n.a.	621.0	1571.2	6.2
2000	776.5	4.3	n.a.	n.a.	n.a.	678.2	2079.4	5.6
avg.								

(c) Results for instances with $ld = 20$.

n	GREEDY		CPLEX			CMSA		
	mean	time	mean	time gap		mean	time	size (%)
200	93.9	< 1.0	86.0	0/0	0.0	86.0	0.8	51.0
400	170.2	< 1.0	151.4	4/5	0.0	151.6	101.7	29.1
600	237.4	< 1.0	210.8	28/29	0.0	212.1	688.4	18.1
800	301.3	< 1.0	267.2	99/102	0.0	269.1	741.8	12.1
1000	365.6	< 1.0	323.1	250/277	0.0	326.2	425.6	11.6
1200	426.3	< 1.0	374.0	568/591	0.0	377.7	772.9	8.2
1400	484.4	1.4	426.6	997/1142	0.0	433.6	609.1	7.3
1600	542.0	2.0	477.1	1331/1640	0.2	484.7	792.4	5.5
1800	598.3	3.1	528.1	2043/2387	0.5	535.2	698.2	5.0
2000	663.2	4.1	680.5	2971/3132	7.1	587.7	518.5	5.2
avg.								

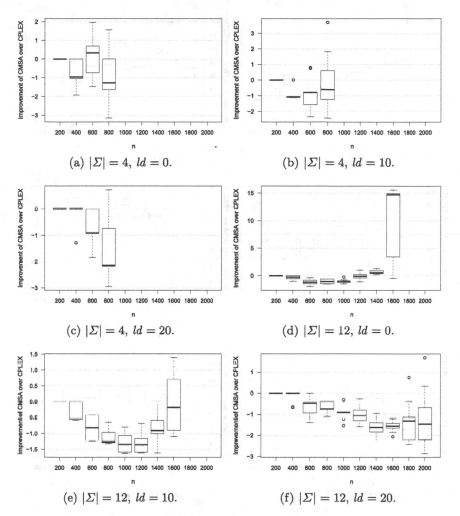

(a) $|\Sigma| = 4$, $ld = 0$.

(b) $|\Sigma| = 4$, $ld = 10$.

(c) $|\Sigma| = 4$, $ld = 20$.

(d) $|\Sigma| = 12$, $ld = 0$.

(e) $|\Sigma| = 12$, $ld = 10$.

(f) $|\Sigma| = 12$, $ld = 20$.

Fig. 1. Improvement of CMSA over CPLEX (in percent). Note that when boxes are missing, CPLEX was not able to provide feasible solutions within the allowed computation time.

– In contrast to CPLEX, CMSA is able to provide feasible solutions for all problem instances. Moreover, CMSA outperforms GREEDY in all cases. In those cases in which CPLEX is able to provide feasible (or even optimal) solutions, CMSA is either competitive, or not much worse than CPLEX. In particular, CMSA is never more than 3 % worse than CPLEX.

In summary, we can state that CMSA is competitive with the application of CPLEX to the original ILP model when the size of the input instances is rather small. The larger the size of the input instances, and the smaller the alphabet size, the greater is—in general—the advantage of CMSA over the other algorithms.

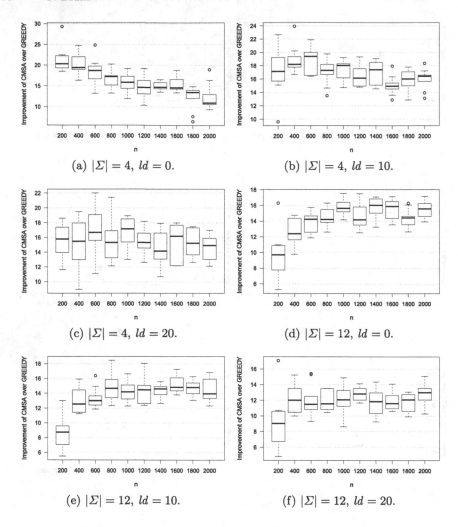

Fig. 2. Improvement of CMSA over GREEDY (in percent).

Finally, we also present the size of the sub-instances that are generated (and maintained) within CMSA in comparison to the size of the original problem instances. These sub-instance sizes are provided in a graphical way in Fig. 3. Note that these graphics show the sub-instance sizes averaged over all instances of the same alphabet size and the same value for ld. In all cases, the x-axis ranges from instances with a small base-length (n) at the left, to instance with a large base-length at the right. Interestingly, when the base-length is rather small, the tackled sub-instances in CMSA are rather large (up to $\approx 55\%$ of the size of the original problem instances). With growing base-length, the size of the tackled sub-instances decreases. The reason for this trend is as follows. As CPLEX is very efficient for problem instances created with rather small

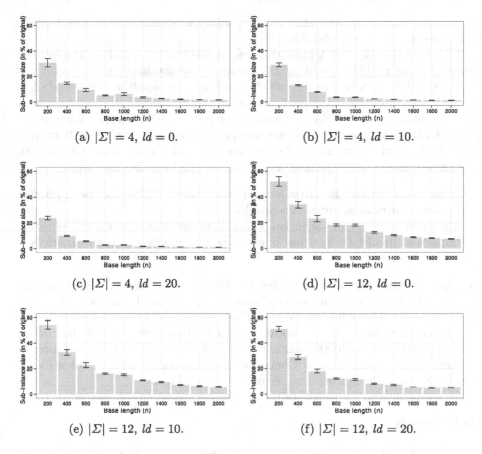

Fig. 3. Graphical presentation of the sizes of the sub-instances in percent with respect to the size of the original problem instances.

base-lengths, the parameter values of CMSA are chosen during the tuning process of irace such that the sub-instance sizes become quite large. On the contrary, with growing base-length, the parameter values chosen during tuning lead to smaller sub-instances, simply because CPLEX is not so efficient anymore when applied to sub-instances that are not much smaller than the original problem instances.

6   Discussion and Future Work

CMSA is a recently proposed, general, algorithm for combinatorial optimization. The algorithm is based on a simple, but seemingly successful, idea: (1) the creation of sub-instances based on merging the solution components found in randomly constructed solutions, and (2) the subsequent solution of these sub-instances by means of an exact solver. In this process, the considered sub-instances undergo dynamic changes caused by adding new solution components at each iteration, and removing existing solution components on the basis of indicators about their usefulness.

In this work, the CMSA algorithm was applied to the unbalanced minimum common string partition problem. The nature of the obtained results, in comparison to CPLEX, is similar to the one observed in earlier applications of CMSA to the minimum common string partition problem and a minimum weight arborescence problem in [3]. CMSA is generally competitive with—or not much worse than—CPLEX for small to medium size problem instances, whereas it outperforms CPLEX with growing problem instances size. In the opinion of the authors, this algorithm is quite appealing, especially for the following reasons:

- Given a constructive heuristic and an exact solver, CMSA can be applied to any combinatorial optimization problem.
- Compared to other metaheuristics, the implementation of CMSA is rather simple and involves only a few lines of code in addition to the heuristic and the exact solver.

Finally, it is important to observe that the idea behind CMSA is related, in some sense, to the basic idea of large neighborhood search (LNS) [11]. However, while LNS uses exact solvers for searching the best solution in a large neighborhood of the current solution, generally obtained by a partial destruction of the current solution, exact solvers in the context of CMSA are applied to sub-instances of the original problem instances. Concerning future work, we plan to indentify the respective strengths and weaknesses of LNS and CMSA.

References

1. Blum, C., Calvo, B.: A matheuristic for the minimum weight rooted arborescence problem. J. Heuristics **21**(4), 479–499 (2015)
2. Blum, C., Lozano, J.A., Pinacho Davidson, P.: Iterative probabilistic tree search for the minimum common string partition problem. In: Blesa, M.J., Blum, C., Voß, S. (eds.) HM 2014. LNCS, vol. 8457, pp. 145–154. Springer, Heidelberg (2014)
3. Blum, C., Pinacho, P., López-Ibáñez, M., Lozano, J.A.: Construct, Merge, Solve & Adapt: A new general algorithm for combinatorial optimization. Comput. Oper. Res. **68**, 75–88 (2016)
4. Bulteau, L., Fertin, G., Komusiewicz, C., Rusu, I.: A fixed-parameter algorithm for minimum common string partition with few duplications. In: Darling, A., Stoye, J. (eds.) WABI 2013. LNCS, vol. 8126, pp. 244–258. Springer, Heidelberg (2013)
5. Chen, X., Zheng, J., Fu, Z., Nan, P., Zhong, Y., Lonardi, S., Jiang, T.: Computing the assignment of orthologous genes via genome rearrangement. In: Proceedings of the Asia Pacific Bioinformatics Conference 2005, pp. 363–378 (2005)
6. Ferdous, S.M., Sohel Rahman, M.: Solving the minimum common string partition problem with the help of ants. In: Tan, Y., Shi, Y., Mo, H. (eds.) ICSI 2013, Part I. LNCS, vol. 7928, pp. 306–313. Springer, Heidelberg (2013)
7. Ferdous, S.M., Sohel Rahman, M.: A MAX-MIN ant colony system for minimum common string partition problem.*CoRR*, abs/1401.4539 (2014). http://arxiv.org/abs/1401.4539

8. Goldstein, A., Kolman, P., Zheng, J.: Minimum common string partition problem: hardness and approximations. In: Fleischer, R., Trippen, G. (eds.) ISAAC 2004. LNCS, vol. 3341, pp. 484–495. Springer, Heidelberg (2004)
9. He, D.: A novel greedy algorithm for the minimum common string partition problem. In: Măndoiu, I.I., Zelikovsky, A. (eds.) ISBRA 2007. LNCS (LNBI), vol. 4463, pp. 441–452. Springer, Heidelberg (2007)
10. López-Ibáñez, M., Dubois-Lacoste, J., Stützle, T., Birattari, M.: The irace package, iterated race for automatic algorithm configuration. Technical report TR/IRIDIA/2011-004, IRIDIA, Université Libre de Bruxelles, Belgium (2011)
11. Pisinger, D., Ropke, S.: Large neighborhood search. In: Gendreau, M., Potvin, J.-Y. (eds.) Handbook of Metaheuristics. International Series in Operations Research & Management Science, vol. 146, pp. 399–419. Springer, US (2010)

Variable Neighbourhood Descent with Memory: A Hybrid Metaheuristic for Supermarket Resupply

Philip Mourdjis[1](✉), Yujie Chen[1], Fiona Polack[1],
Peter Cowling[1], and Martin Robinson[2]

[1] Department of Computer Science, University of York, York, UK
{pjm515,yc1005,fiona.polack,peter.cowling}@york.ac.uk
[2] Transfaction Ltd., Cambridge, UK
martin.robinson@transfaction.com

Abstract. Supermarket supply chains represent an area in which optimisation of vehicle routes and scheduling can lead to huge cost and environmental savings. As just-in-time ordering practices become more common, traditionally fixed resupply routes and schedules are increasingly unable to meet the demands of the supermarkets. Instead, we model this as a dynamic pickup and delivery problem with soft time windows (PDPSTW). We present the variable neighbourhood descent with memory (VNDM) hybrid metaheuristic (HM) and compare its performance against Q-learning (QL), binary exponential back off (BEBO) and random descent (RD) hyperheuristics on published benchmark and real-world instances of the PDPSTW. We find that VNDM consistently generates the highest quality solutions, with the fewest routes or shortest distances, amongst the methods tested. It is capable of finding the best known solutions to 55 of 176 published benchmarks as well as producing the best results on our real-world data set, supplied by Transfaction Ltd.

1 Introduction

There are over ten thousand grocery stores or supermarkets in the UK [29] and each of these requires regular resupplies of produce. With increasingly heavy competition in the market, just-in-time stock management is increasingly important to meet customer demand whilst managing food waste. Traditional, fixed resupply routes and schedules are unable to cope with these demands. We propose a pickup and delivery problem with soft time windows (PDPSTW) as a more suitable way of scheduling supermarket resupplies.

In this paper, we introduce the variable neighbourhood descent with memory (VNDM) to solve this problem and compare its performance against QL, BEBO and RD hyperheuristics on both benchmark instances and real-world data. We show that VNDM is an effective approach particularly for clustered benchmark instances and real-world data sets. We also investigate the landscape of potential solutions for both benchmark and real-world data and show that minimising the number of routes in a solution does not always produce an improvement in cost, often the most important factor for supermarkets and delivery companies.

© Springer International Publishing Switzerland 2016
M.J. Blesa et al. (Eds.): HM 2016, LNCS 9668, pp. 32–46, 2016.
DOI: 10.1007/978-3-319-39636-1_3

2   Related Work

The pickup and delivery problem with time windows (PDPTW) is an NP-hard combinatorial optimisation problem that involves routing vehicles to service requests from pickup to delivery locations [19]. It differs from vehicle routing problems (VRP) in that cargo is loaded at numerous pickup locations instead of a vehicle's depot. Research on real-world PDPTWs usually concentrates on static models of small scale dial-a-ride problems, such as taxi routing and ride sharing schemes [31]. Dynamic problems [11] typically do not compare their techniques on benchmark instances. Variants to the PDPTW include imposing constraints on: the number of vehicles used, time windows on requests, capacities and number of depots. Our problem of supermarket resupply is based on widely accepted mathematical models of the PDPTW by Li and Lim [19] and Desaulniers et al. [10]; with additional constraints specific to our problem.

Exact algorithms e.g. [12,14,33] have been used to solve PDPTWs with tens of requests but do not scale to real world problems with thousands of requests. Recently, heuristic and hyperheuristic [9] approaches have become popular. These approaches cannot guarantee an optimal solution, but often produce very near optimal solutions far quicker than exact methods. A good overview of exact and heuristic methods for VRP can be found in [18]. Meta-heuristics deliver a mechanism that helps a search escape local optima. Successful application of meta-heuristics including variable neighbourhood search (VNS) and Tabu search (TS) for both VRP and PDPTW can be found in [8,22]. Hyperheuristics define a high level set of rules that govern when to use local search operators (LSOs) either deterministically or based on previous performance. In comparison, HMs typically make greater use of domain knowledge [4] and utilize the advantages of a set of different search methods. HMs e.g. [6,20,21,23,26], have proved successful when applied to the VRP but have seen limited applicability to the PDPTW. Hybridisation of VNS and TS has proved beneficial for VRPs; Paraskevopoulos et al. [21] presents a reactive variable neighbourhood tabu search (reVNTS) for the heterogeneous VRP and Belhaiza et al. [1] presents a hybrid variable neighbourhood tabu search (HVNTS) for the VRP with time windows, the method presented in this paper has similarities with these methods and is discussed in Sect. 4.

Benchmark instances of the PDPTW (from Li and Lim [19]) will be used to compare VNDM against other methods and the state of the art. These benchmarks are more tightly constrained than our supermarket resupply problem and use a different objective function but are otherwise similar enough for comparisons to be made. The benchmarks have been investigated by many academic [2,3,15–17,19,27] and corporate [24,30] researchers. The best published results are kept up to date at sintef.no/pdptw.

3   Problem Definition

In the supermarket resupply problem a set of vehicles based at a single depot must service a set of consignments, each comprised of a pickup and a corresponding delivery location. An un-/loading time is associated with each pickup and delivery location along with a time windows specifying the earliest and latest times that service may begin. Arrival before a time window results in waiting for the earliest service time before un-/loading may commence. Load is added to a vehicle at pickup locations and removed at corresponding delivery locations in the same quantity. A vehicle may carry any number of consignments simultaneously, as long as capacity constraints are not violated. All vehicle routes begin and end at the depot and are empty, having zero load, at both these points. Additionally, a maximum time limit is placed on every vehicle route, calculated from the combined time required to traverse all locations in the route, un-/load at each location and any waiting time incurred.

The mathematical model for the PDPTW can be found in Li and Lim [19]. The main differences are that the supermarket resupply problem is a dynamic system where not all orders are known a priori, and that time windows are "soft", if some locations are arrived at late a delay penalty is incurred (increasing over time) instead of a route being infeasible. The objective, instead of minimising the number of vehicles used and then total distance travelled, is to minimise cost as a function of: fuel and maintenance costs dependent on distance travelled and load; time costs of drivers pay and delay penalties. We do not consider idle driver time outside of a route, for example if a route is short then the driver is only paid for the time on route (not for number of days worked).

4   Solution Method

A solution has been developed to handle both static instances such as the benchmark problems and our real-world data which we treat as a dynamic problem. In both cases consignments are first inserted greedily into a schedule in first come, first served fashion (placed on an existing vehicle's route which minimises cost if possible, else creating a new route), optimisation is then performed both inside and between the resultant routes using one of a number of methods: Variable Neighbourhood Descent with Memory (VNDM, Sect. 4.3); Q-learning (QL, Sect. 4.4); Binary exponential back off (BEBO, Sect. 4.5); Random Descent (RD, Sect. 4.6). In the dynamic case this optimisation is performed multiple times (between the arrival of subsequent orders) and an internal representation of time is held so that consignments which would have taken place if the system was being used live are not re-scheduled during later optimisation. Each of the methods above makes use of a set of local search operators (LSOs, Sect. 4.1) that allow small changes to the existing schedule to be analysed and adopted if improvements are found. A re-initialisation step, used to escape local optima (Sect. 4.2), is also common in these methods, and is performed after an internally defined number of iterations without improvement. The best solution found at any point during execution is stored and reported upon termination.

4.1   Low Level Heuristics

LSOs for the PDPSTW have been drawn from similar problems [5,7,10,13,28] and chosen to cover a wide range of potential variations from an existing schedule. Since a pickup request must occur before its delivery requests, reversing a section of a route is likely to result in an infeasible solution. Time windows are also usually tight enough that reversal would mean that one or more requests would be significantly delayed. LSOs relying on partial route inversions such as GENI [13] and iCROSS [5] cannot work well without substantial alteration, so standard relocate and chain exchange operators have been preferred. Specifically we use: Exchange; Exchange Chain (Cross exchange [5]); Relocate and Relocate Chain operators that may act either within a route or between two routes. Exchange operators swap the positions of two (chains of) consignments while relocate operators move only a single (chain of) consignments. If an individual consignment is moved, any requests nested within it are not moved with it unless they fall in the same chain of consignments. This provides a means to undo nested consignments - if it can provide an improvement elsewhere. Chains have a fixed maximum length of 5 consignments.

Algorithm 1. VNDM

1: **function** VNDM(Schedule s)	
2: $s* \leftarrow s$	
3: **repeat**	
4: **for all** l in LSOs	▷ Section 4.1
5: **repeat**	
6: First Improvement (s, l)	▷ Algorithm 2
7: **if** found improvement	
8: Update s with improvement	
9: **if** s better than $s*$	
10: $s* \leftarrow s$	
11: **until** no improvement for l	
12: Shake s	▷ Section 4.2
13: **until** current time $\geq$ time limit	
14: **return** $s*$	

4.2   Random Re-initialisation (Shaking)

All the methods compared in this paper benefit from random re-initialisation, a large, arbitrary permutation made to the current solution once no LSOs can produce any further improvement. It can be thought of as a random re-start of the search from a different initial solution, or a more extreme example of a shake operation that may be used in VNS. For the PDPSTW, this step consists of removing a random number of routes (drawn uniformly between 1 and the number of routes in the solution) and a random number of additional customers (drawn between 1 and the number of customers left in the solution). The combined list of all removed customers is then re-inserted into the remaining routes, creating new additional routes if required. The scale of the destruction of the original solution normally results in a substantially different solution from which to restart the search, though we do not guarantee it is unique.

4.3   Variable Neighbourhood Descent with Memory

VNDM is a variant of VNS with a strong bias towards exploitation, using a first improvement descent strategy (described in Algorithm 1). Shaking (the use of a random perturbation on the solution to encourage diversification) is only performed once no LSOs are capable of producing improving moves, in contrast to traditional VNS where it is used at every neighbourhood. Additionally, VNDM stores a tabu memory of information on which routes have been analysed by which LSOs, described in Algorithm 2. This hybridisation is inspired by the related reVNTS [21] and HVNTS [1]. In contrast, HVNTS stores recently seen solutions and distinguishes between large and small moves in its neighbourhood structure. ReVNTS uses tabu search to find a local optimum within each neighbourhood of a VNS; additional features are learnt to control the use of LSOs.

Algorithm 2. First Improvement

Precondition: Tabu db storing route and LSO IDs
Precondition: RouteList rl sorted by fitness descending
 1: **function** FIRST IMPROVEMENT(Schedule s, LSO l)
 2:     **repeat**
 3:         selectedRoutes $sr \leftarrow rl$.GetNextRoutes(l,s)
 4:         **if** (sr &l) not in db
 5:             moves $\leftarrow l$.GetMoves(sr)
 6:             **for all** move m in moves
 7:                 fitnessDelta $d \leftarrow l$.Test(m)
 8:                 **if** $d < 0$ ▷ improvement
 9:                     db.Remove(sr)   ▷ other improvements now possible
10:                     **return** sr, m &d
11:             db.Add(sr, l)   ▷ no improvement found, add to memory
12:         **else**
13:             do nothing   ▷ tried before and found no improvement
14:         rl.Remove(sr)
15:     **until** no more routes
16:     **return** null   ▷ No improvement possible using l

In Algorithm 2, routeList rl contains all k-element subsets of route ids in the schedule where k is the number of routes required by LSO l. rl is ordered by the summed fitness of the routes in each k-element subset, descending (lower fitness is better). rl.GetNextRoutes(l,s) returns the selected routes sr (the first k-element subset in rl). This is then removed from rl so it is only chosen once per LSO l. $l.GetMoves(sr)$ generates all potential moves M for a given LSO l on sr. $l.Test(m)$ generates the difference in fitness for a given move $m \in M$ with LSO l. If the fitness delta d is less than 0 (an improvement), the selected routes, move and fitness delta are returned. The move is then applied to the schedule, updating fitnesses for individual routes as appropriate. The memory is updated to remove the altered routes from all LSOs as this move may have enabled changes that were not previously possible.

4.4   Q-Learning Selection Based Hyperheuristic (QL)

QL [32] is a learning hyperheuristic that shares a number of similarities with the choice function of Cowling et al. [9]. QL attempts to learn good sequences of LSOs, these are stored in a Q-state dictionary which maps sequences of n LSOs to Q-value. At each iteration, QL identifies sequences from the dictionary that start with the most recently used $n - 1$ LSOs. The next LSO to try is chosen based on a roulette selection over these entries Q-values. The Q-values in the dictionary are updated using the function [32]:

$$Q(s,a) = Q(s,a) + \alpha \left[r + \left(\gamma \max_{a'} Q(s',a') \right) - Q(s,a) \right] \qquad (1)$$

where s is the current $1, \ldots, n - 1$ sequence of LSOs, a is the next LSO to use and s' is the resultant sequence, after this operator is used. The reward r is set to the improvement produced by the operator, divided by the time taken to find it or to half the smallest observed reward if no improvement is found. α, the learning rate, γ, the discount factor, and n, the length of LSO sequences to store, are parameters. Traditional QL allows non-improving moves. However, since our problem has a very limited set of improving moves, we have adapted it to only accept moves which result in better solutions.

4.5   Binary Exponential Back-off Tabu Search

BEBO [25] is a tabu based hyperheuristic derived from methods used to avoid packet collision in communications systems. A tabu list is stored along with a backoff value for each LSO. At each iteration, all non-tabued LSOs are tested. These are then categorised as good or bad. A "Bad" LSO may be any non-improving move, or may be defined as being one of the worst x % of LSOs. "Good" LSOs have their backoff set to backoff-min. "Bad" LSOs have their backoff set to backoff$^2 + 1$. A tabu tenure is then chosen for each LSO, randomly between backoff-min and its backoff.

4.6   Random Descent

The final approach is a simple RD heuristic. At each iteration, a random LSO is chosen and tested. If the LSO produces a better solution, it is accepted; otherwise it is discarded. A parameter is used to control the number of non-improving iterations that are allowed before the solution is re-initialised.

5   Computational Experiments

We first perform parameter tuning (Sect. 5.1) and secondly compare the performance of the four methods presented in Sect. 4 on all 100, 200 and 400 customer instances of the Li and Lim [19] benchmarks (Sect. 5.2), In these we consider hard time windows and no LIFO constraint such that direct comparisons may be made to best known solutions. Finally we compare performance on a set of real-world instances (Sect. 5.3) with our additional constraints.

The solution methods are coded in single threaded C$^\sharp$ and distributed over a heterogeneous cluster of Intel Xeon based servers totalling 72 cores and 120 GB of RAM. All methods are given 5, 10 or 20 min of CPU time based on problem size (100, 200 or 400 customers respectively) and each is repeated 10 times. The results presented in this paper thus represent over 1200 h of CPU time.

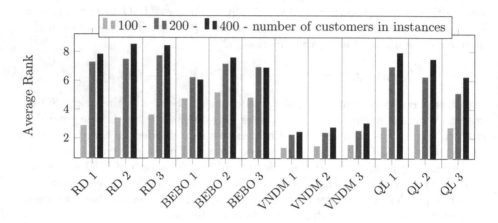

Fig. 1. Average rank of heuristics on different sized problems (lower is better).

5.1   Parameter Tuning

Each method is tested using three sets of parameters. For RD there is only one parameter, the number of iterations before re-initialisation. In RD 1, 2 and 3 it is set at 250, 500 and 1000 respectively. The parameter we modify for BEBO controls the number of LSOs that are backed off at any stage of the search. The three values chosen for BEBO 1, 2 and 3 represent backing off all but the best solution, or all solutions more than 5 % or 10 % worse than the best solution (see Sect. 4.5). VNDM has two parameters, controlling the number of iterations (a) without improvement before re-initialisation and (b) before reverting to the previous best solution. These are set as (50, 125), (100, 250) and (200, 500) respectively for VND 1, 2 and 3. For QL, we investigate changes to the learning function (Eq. 1) by setting α and γ to (0.25, 0.75), (0.5, 0.5) and (0.75, 0.25) respectively for QL 1, 2 and 3.

The heuristics are ranked for each instance, using their best result over 10 runs, and are given a score from one (best) to twelve (worst) where ties score equally low ranks. Figure 1 shows the average rank for each of the twelve methods for the 100, 200 and 400 customer instances. The 100 customer instances have a generally lower average rank because, in many of these cases the best known solution is found by several methods which are then given equal rank one. An interesting point is that BEBO is weaker on smaller data sets but better on larger data sets compared to both QL and RD, suggesting that the overhead involved in trialling sets of LSOs is not worth the effort on small instances but is able to find better solutions on larger problems. VNDM consistently outperforms its competitors whilst demonstrating robustness to changes in parameter settings.

5.2   Comparison on Benchmark Instances

The Li and Lim benchmarks [19] are split into three groups characterised by the spatial characteristics of the problem: Random instances (LRx-y-z) have customer locations that are spread uniformly randomly across space; clustered instances (LCx-y-z) have customer locations that are tightly grouped into a number of distinct clusters; and mixed instances (LRCx-y-z) have a mix of both random and clustered locations. For each instance, x can be either 1 (tight time windows) or 2 (lax time windows); y represents the number of customers in the instance (divided by 100) and z is the instance id.

Presented in Tables 1 and 2 are the best results of 10 repeats for each heuristic, obtained using the best performing parameters identified in Sect. 5.1, for the 100 and 400 customer clustered instances respectively. For each method, the r and d columns are the number of routes and distance for the best observed run. The gap column records the difference between our best solution and the best known solutions of [2, 3, 15–17, 19, 24, 27, 30] as reported at sintef.no/pdptw. Where our best solution has the same number of routes but is longer, the gap records this difference as a percentage. Where our solution has n extra routes, the gap is nr. We highlight in **bold** the best solutions found out of our results.

The four hyperheuristics perform very similarly on the 100 customer instances, finding the best known solutions in many cases: these instances are shown with their name in **bold**. In cases where the best known solution is not found, VNDM is the best or joint best of the methods tested and produces results within 0.3 % or 1 route of the best known solution[1].

Table 2 shows the results of the 400 customer clustered instances. It is clear that VNDM is the strongest method we compare, matching best known solutions in many cases. All our methods have difficulties with the LC2 data set due to looser time constraints resulting in a much larger quantity of feasible solutions. VNDM still produces results that either have fewer routes or, for instances with the same number of routes, solutions that are on average around 10 % shorter than the other three approaches. However, VNDM often produces more routes in comparison to best known solutions.

---

[1] Full results are available at www.cs.york.ac.uk/~philm.

Table 1. 100 customers clustered. For explanation see text.

Name	RD		BEBO		VNDM		QL		Gap
	r	d	r	d	r	d	r	d	
LC1-1-1	**10**	**828.94**	**10**	**828.94**	**10**	**828.94**	**10**	**828.94**	0%
LC1-1-2	**10**	**828.94**	**10**	**828.94**	**10**	**828.94**	**10**	**828.94**	0%
LC1-1-3	9	1072.83	9	1082.18	**9**	**1038.35**	9	1048.40	0.29%
LC1-1-4	9	904.10	9	993.98	**9**	**861.95**	9	876.88	0.23%
LC1-1-5	**10**	**828.94**	**10**	**828.94**	**10**	**828.94**	**10**	**828.94**	0%
LC1-1-6	**10**	**828.94**	**10**	**828.94**	**10**	**828.94**	**10**	**828.94**	0%
LC1-1-7	**10**	**828.94**	**10**	**828.94**	**10**	**828.94**	**10**	**828.94**	0%
LC1-1-8	**10**	**826.44**	**10**	**826.44**	**10**	**826.44**	**10**	**826.44**	0%
LC1-1-9	**10**	**827.82**	10	882.86	**10**	**827.82**	**10**	**827.82**	1r
LC2-1-1	**3**	**591.56**	**3**	**591.56**	**3**	**591.56**	**3**	**591.56**	0%
LC2-1-2	**3**	**591.56**	**3**	**591.56**	**3**	**591.56**	**3**	**591.56**	0%
LC2-1-3	**3**	**591.17**	3	772.52	**3**	**591.17**	**3**	**591.17**	0%
LC2-1-4	3	676.03	3	614.65	**3**	**590.60**	3	652.95	0%
LC2-1-5	**3**	**588.88**	**3**	**588.88**	**3**	**588.88**	**3**	**588.88**	0%
LC2-1-6	**3**	**588.49**	**3**	**588.49**	**3**	**588.49**	**3**	**588.49**	0%
LC2-1-7	**3**	**588.29**	3	606.10	**3**	**588.29**	**3**	**588.29**	0%
LC2-1-8	3	591.39	3	594.69	**3**	**588.32**	**3**	**588.32**	0%

Closer investigation of the results for various instances shows that a wide array of solutions is created, each of which is subtly different. Since all our heuristics are based on first improvement and have ample time to converge, this can be attributed not to time constraints but to the nature of the problem itself. The solution landscape for PDPTWs is not smooth and contains many local optima, making it difficult for heuristics to converge on the same result. To visualise this, we plot all of our results in two dimensions, showing the number of routes against schedule distance. Figure 2 presents solution space maps generated in this way for four representative instances.

Figure 2 highlights an interesting finding in the benchmark data sets. In the clustered (LC) data sets there is a clear trend between the number of routes and the total distance of a solution. In the random data set, however, these aspects are not closely correlated, so using the total number of routes as the main objective does not seem to be appropriate. We note also that, as expected, the distance is greater in the random scenarios, however, the number of routes is lower, probably due to the looser time window constraints and smaller service times in these problems. Koning [17] notes that relaxing the hard time window constraints, applied in the Li and Lim [19] benchmarks, produces notably shorter routes with only minor delays, which may be preferable in the real world.

Table 2. 400 customers clustered. For explanation see text.

Name	RD		BEBO		VNDM		QL		Gap
	r	d	r	d	r	d	r	d	
LC1-4-1	**40**	**7152.06**	40	7208.31	**40**	**7152.06**	**40**	**7152.06**	0 %
LC1-4-2	40	7184.42	40	7491.64	**40**	**7170.60**	40	7235.69	2r
LC1-4-3	37	8089.33	37	8684.88	**37**	**7871.19**	37	8383.98	4r
LC1-4-4	32	8328.28	32	8544.82	**32**	**7403.17**	32	7748.67	2r
LC1-4-5	**40**	**7150.00**	40	7150.00	**40**	**7150.00**	**40**	**7150.00**	0 %
LC1-4-6	**40**	**7154.02**	40	7237.16	**40**	**7154.02**	40	7170.01	0 %
LC1-4-7	41	7542.55	42	8734.36	**40**	**7149.44**	41	7435.92	0 %
LC1-4-8	**39**	**7111.16**	40	7706.57	39	7179.98	**39**	**7111.16**	0 %
LC1-4-9	38	8197.97	38	8390.38	**37**	**7819.79**	39	8479.35	1r
LC1-4-10	38	7940.02	37	8016.53	**37**	**7670.50**	37	7990.68	2r
LC2-4-1	14	6824.82	**12**	**4116.33**	12	4116.33	13	5444.85	0 %
LC2-4-2	14	9135.06	13	5108.89	**13**	**4844.74**	14	7999.00	1r
LC2-4-3	13	7145.52	13	5967.34	**12**	**5364.88**	13	6375.44	1r
LC2-4-4	13	7727.34	12	6193.68	**12**	**5766.83**	13	7311.49	35 %
LC2-4-5	15	8612.69	13	5243.16	**13**	**4717.13**	14	6886.47	1r
LC2-4-6	14	7560.98	13	4936.46	**13**	**4721.75**	14	7125.17	1r
LC2-4-7	14	8312.98	14	5882.84	**13**	**4616.22**	14	7542.35	2r
LC2-4-8	14	7883.71	13	5456.19	**13**	**4523.78**	14	7582.07	1r
LC2-4-9	14	7770.37	13	6334.76	**13**	**5419.32**	14	7450.32	1r
LC2-4-10	14	7867.45	**13**	**4655.07**	13	4737.62	13	6330.36	1r

5.3 Performance on Real World Data

Our real-world data set comprises 387 consignments pairs spread over the south east of the UK. The data is provided by a large supermarket chain and comprises resupply deliveries for a one-week period. Due to the distribution of store locations, a larger number of routes is required for full service, in comparison to similarly sized benchmark problems. In addition, for the real-world problem we simulate consignments arriving during scheduling in a dynamic fashion, allow late arrival at locations using "soft" end of time windows and we use a cost based model as the objective rather than the number of routes and distance (described in Sect. 3). The real-world data set is run with the best parameter settings of each method from Sect. 5.1 for 40 min and repeated 100 times.

Figure 3 shows the average performance over CPU time. From discussions with our industrial partner, current manual scheduling procedures are most closely approximated by our initial greedy insertion procedure, the cost of which is represented by the starting figure of around £33,500 in this example data set. Clearly, utilising any of the methods in this paper results in large savings to

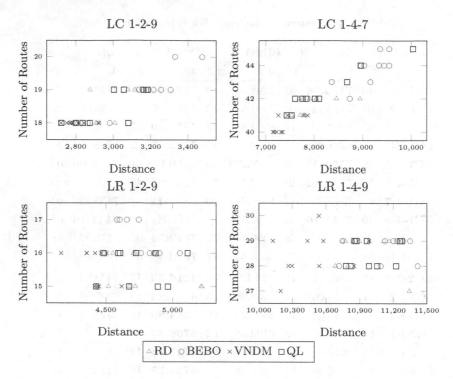

Fig. 2. Number of routes versus cost for clustered (LC 1-2-9 and LC 1-4-7) and random (LR 1-2-9 and LR 1-4-9) benchmark instances.

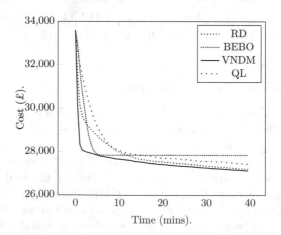

Fig. 3. Average performance over CPU time for a real data set, note that BEBO converges after about 5 min and then no longer produces any improvements.

delivery cost. On average, and over any amount of running time, VNDM can find better solutions for the real-world problem than the alternatives tested here and

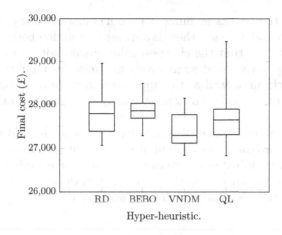

Fig. 4. Distribution of final cost across 100 runs.

is relatively simple to implement. We can also see that BEBO produces results that are competitive if run-times are kept below 10 min but that it gets stuck at a worse level than other approaches after this.

Figure 4 shows the distribution of final cost achieved by each hyperheuristic across the runs. There is considerable overlap in the plots, though VNDM can be seen to outperform the alternatives, having lower minimum, maximum and quartile costs. It is also clear that BEBO is unsuitable for the PDPSTW, as it produces the highest average final cost and is generally outperformed by RD. QL looks slightly better in these results than in Fig. 3 as its best and many of its results are competitive, but it also produces the most expensive schedules, hurting its average.

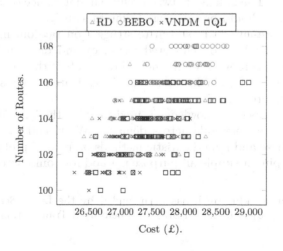

Fig. 5. Number of routes vs cost for real data set.

Figure 5 plots the results as number of routes against distance. We can see that, for our real-world data set, there is a strong correlation between the number of routes and cost, but that the cheapest solutions do not necessarily have the fewest routes. Approaches that aggressively minimise for number of routes may not be able to perform as well as the approaches here from an operational cost perspective. The real-world results most closely resemble the LC 1-4-z benchmark instances.

Though we have only provided results for a single real-world data set, since the VNDM metaheuristic proved competitive with state of the art approaches on many small benchmark instances, we can say with some confidence that VNDM is a suitable choice for our real-world problem, even though it has undergone no parameter tuning to specifically fit the real-world data.

6   Conclusions

We have presented the supermarket resupply problem and modelled it as a dynamic PDPSTW. We introduced the VNDM hybrid metaheuristic to solve this problem and shown it to be competitive with best known solutions for small benchmark instances of the PDPTW (by Li and Lim [19]). We have shown that, in limited CPU time, VNDM outperforms BEBO, QL and RD on many of the 100, 200 and 400 customer benchmark instances and on real-world data. This result is demonstrated with a variety of different parameter settings and is not overfit to either benchmark instances or the real-world data set.

For many of the random and some of the clustered instances in the Li and Lim benchmarks, shorter solutions are possible if more routes are used. We have shown that the traditional PDPTW objective of minimising the number of routes in priority to total distance does not always produce the cheapest solutions in a real-world problem. The balance between vehicle maintenance costs and distance based running costs should be considered simultaneously.

The methods presented are shown to struggle on random instances of the PDPTW, performing best on clustered instances and data such as our real-world problem. In future work we intend to investigate if this conclusion is valid for other real-world PDPSTW instances, and investigate methods of scaling this approach to larger problem sizes.

Though domain specific knowledge is used to guide the VNDM, the majority of the method is transportable across domains. We intend to investigate the performance of these and hyperheuristic methods on related problem areas with the aim of developing a simple all purpose method for combinatorial problems.

Acknowledgements. This work has been funded by the Large Scale Complex IT Systems (LSCITS) project of the EPSRC with help from Transfaction Ltd.

References

1. Belhaiza, S., Hansen, P., Laporte, G.: A hybrid variable neighborhood tabu search heuristic for the vehicle routing problem with multiple time windows. Comput. Oper. Res. **52**(part B), 269–281 (2013)
2. Bent, R., Van Hentenryck, P.: A two-stage hybrid algorithm for pickup and delivery vehicle routing problems with time windows. In: Rossi, F. (ed.) CP 2003. LNCS, vol. 2833, pp. 123–137. Springer, Heidelberg (2003)
3. Blocho, M.: A parallel algorithm for minimizing the fleet size in the pickup and delivery problem with time windows. In: Proceedings of the 22nd European MPI Users' Group Meeting, pp. 20–21. ACM (2015)
4. Blum, C., Puchinger, J., Raidl, G.R., Roli, A.: Hybrid metaheuristics in combinatorial optimization: a survey. Appl. Soft Comput. **11**(6), 4135–4151 (2011)
5. Bräysy, O.: A reactive variable neighborhood search for the vehicle-routing problem with time windows. INFORMS J. Comput. **15**(4), 347–368 (2003)
6. Carić, T., Fosin, J., Galić, A., Gold, H., Reinholz, A.: Empirical analysis of two different metaheuristics for real-world vehicle routing problems. In: Bartz-Beielstein, T., Blesa Aguilera, M.J., Blum, C., Naujoks, B., Roli, A., Rudolph, G., Sampels, M. (eds.) HCI/ICCV 2007. LNCS, vol. 4771, pp. 31–44. Springer, Heidelberg (2007)
7. Cherkesly, M., Desaulniers, G., Laporte, G.: Branch-price-and-cut algorithms for the pickup and delivery problem with time windows and LIFO loading. Comput. Oper. Res. **62**(1), 23–35 (2015)
8. Cordeau, J.F., Laporte, G.: A tabu search heuristic for the static multi-vehicle dial-a-ride problem. Transp. Res. Part B Methodol. **37**, 579–594 (2003)
9. Cowling, P.I., Kendall, G., Soubeiga, E.: A hyperheuristic approach to scheduling a sales summit. In: Burke, E., Erben, W. (eds.) PATAT 2000. LNCS, vol. 2079, pp. 176–190. Springer, Heidelberg (2001)
10. Desaulniers, G., Desrosiers, J., Solomon, M.M., Erdmann, A., Soumis, F.: VRP with pickup and delivery. In: Toth, P., Vigo, D. (eds.) The Vehicle Routing Problem, pp. 225–242. SIAM, Philadelphia (2002)
11. Dorer, K., Calisti, M.: An adaptive approach to dynamic transport optimization. In: Klugl, F., Bazzan, A., Ossowski, S. (eds.) Applications of Agent Technology in Traffic and Transportation. Whitestein Series in Software Agent Technologies, pp. 33–49. Birkhäuser, Basel (2005)
12. Dumas, Y., Desrosiers, J., Soumis, F.: The pickup and delivery problem with time windows. Eur. J. Oper. Res. **54**(1), 7–22 (1991)
13. Gendreau, M., Hertz, A., Laporte, G.: New insertion and post optimization procedures for the traveling salesman problem. Oper. Res. **40**(6), 1086–1095 (1992)
14. Gschwind, T., Irnich, S., Mainz, D.: Effective Handling of Dynamic Time Windows and Synchronization with Precedences for Exact Vehicle Routing. Technical report, Johannes Gutenberg University Mainz, Mainz, Germany. Retrieved from (2012). http://logistik.bwl.uni-mainz.de/
15. Hasle, G., Lie, K.A., Quak, E.: Geometric Modelling, Numerical Simuation, and Optimization. Applied Mathematics at SINTEF, vol. 54. Springer, Heidelberg (2007)
16. Hosny, M.I.: Investigating Heuristic and Meta-Heuristic Algorithms for Solving Pickup and Delivery Problems Manar Ibrahim Hosny School of Computer Science & Informatics. Ph.D. thesis, Cardiff University (2010)
17. Koning, D.: Using Column Generation for the Pickup and Delivery Problem with Disturbances. Masters thesis, Universiteit Utrecht (2011)

18. Laporte, G.: Fifty years of vehicle routing. Transp. Sci. **43**(4), 408–416 (2009)
19. Li, H., Lim, A.: A metaheuristic for the pickup and delivery problem with time windows. In: Tools with Artificial Intelligence, pp. 160–167. IEEE (2001)
20. Ostertag, A., Doerner, K.F., Hartl, R.F.: A variable neighborhood search integrated in the POPMUSIC framework for solving large scale vehicle routing problems. In: Blesa, M.J., Blum, C., Cotta, C., Fernández, A.J., Gallardo, J.E., Roli, A., Sampels, M. (eds.) HM 2008. LNCS, vol. 5296, pp. 29–42. Springer, Heidelberg (2008)
21. Paraskevopoulos, D.C., Repoussis, P.P., Tarantilis, C.D., Ioannou, G., Prastacos, G.P.: A reactive variable neighborhood tabu search for the heterogeneous fleet vehicle routing problem with time windows. J. Heuristics **14**(5), 425–455 (2008)
22. Parragh, S.N., Doerner, K.F., Hartl, R.F.: Variable neighborhood search for the dial-a-ride problem. Comput. Oper. Res. **37**(6), 1129–1138 (2009)
23. Pirkwieser, S., Raidl, G.R.: Multiple variable neighborhood search enriched with ILP techniques for the periodic vehicle routing problem with time windows. In: Blesa, M.J., Blum, C., Di Gaspero, L., Roli, A., Sampels, M., Schaerf, A. (eds.) HM 2009. LNCS, vol. 5818, pp. 45–59. Springer, Heidelberg (2009)
24. Quintiq: PDPTW World Records (2015). http://www.quintiq.com/optimization/pdptw-world-records.html
25. Remde, S., Cowling, P.I., Dahal, K., Colledge, N., Selensky, E.: An empirical study of hyperheuristics for managing very large sets of low level heuristics. J. Oper. Res. Soc. **63**(3), 392–405 (2011)
26. Repoussis, P.P., Paraskevopoulos, D.C., Tarantilis, C.D., Ioannou, G.: A reactive greedy randomized variable neighborhood tabu search for the vehicle routing problem with time windows. In: Almeida, F., Blesa Aguilera, M.J., Blum, C., Moreno Vega, J.M., Pérez Pérez, M., Roli, A., Sampels, M. (eds.) HM 2006. LNCS, vol. 4030, pp. 124–138. Springer, Heidelberg (2006)
27. Ropke, S., Pisinger, D.: An adaptive large neighborhood search heuristic for the pickup and delivery problem with time windows. Transp. Sci. **40**(4), 455–472 (2005)
28. Savelsbergh, M.W.P.: The vehicle routing problem with time windows: minimizing route duration. INFORMS J. Comput. **4**(2), 146–154 (1992)
29. Statistica: Number of stores of leading grocery retailers in the United Kingdom (UK) as of (2013). http://www.statista.com/statistics/299155/number-of-stores-of-grocery-retailers-supermarkets-united-kingdom-uk/
30. TetraSoft, A.: MapBooking Algoritm for Pickup and Delivery Solutions with Time Windows and Capacity restraints. (2003). http://www.tetrasoft.dk/english-info/
31. Toth, P., Vigo, D.: Heuristic algorithms for the handicapped persons transportation problem. Transp. Sci. **31**(1), 60–71 (1997)
32. Watkins, C.J.C.H., Dayan, P.: Q-learning. Mach. Learn. **8**(3–4), 279–292 (1992)
33. Xu, H., Chen, Z.L., Rajagopal, S., Arunapuram, S.: Solving a practical pickup and delivery problem. Transp. Sci. **37**(3), 347–364 (2003)

Hybridization as Cooperative Parallelism for the Quadratic Assignment Problem

Danny Munera[1], Daniel Diaz[1(✉)], and Salvador Abreu[1,2]

[1] University of Paris 1-Sorbonne/CRI, Paris, France
danny.munera@malix.univ-paris1.fr, daniel.diaz@univ-paris1.fr
[2] Universidade de Évora/LISP, Évora, Portugal
spa@di.uevora.pt

Abstract. The Quadratic Assignment Problem is at the core of several real-life applications. Finding an optimal assignment is computationally very difficult, for many useful instances. The best results are obtained with hybrid heuristics, which result in complex solvers. We propose an alternate solution where hybridization is obtain by means of parallelism and cooperation between simple single-heuristic solvers. We present experimental evidence that this approach is very efficient and can effectively solve a wide variety of hard problems, often surpassing state-of-the-art systems.

Keywords: QAP · Heuristics · Parallelism · Cooperation · Hybridization · Portfolio

1 Introduction

The Quadratic Assignment Problem (QAP) was introduced in 1957 by Koopmans and Beckmann [1] as a model of a facilities location problem. This problem consists in assigning a set of n facilities to a set of n specific locations minimizing the cost associated with the *flows* of items among facilities and the *distance* between them. This combinatorial optimization problem has m any other real-life applications: scheduling, electronic chipset layout and wiring, process communications, turbine runner balancing, data center network topology, to cite but a few [2,3]. This problem is known to be NP-hard and finding effective algorithms to solve it has attracted a lot of research in recent years. To tackle problems of medium or large size ($n > 30$) one must resort to incomplete methods which are designed to quickly provide good, albeit potentially sub-optimal, solutions. This is the case of *metaheuristics*. Since the mid-1980s several metaheuristics have been successfully applied to the QAP: tabu search, simulated annealing, genetic algorithms, GRASP, ant-colonies [3]. For solving the hardest instances, the current trend is to specialize existing heuristics [4,5] often by *combining* different metaheuristics (*hybrid procedures*) [6,7] and/or to resort on parallelism [8,9].

We recently proposed a sequential Extremal Optimization (EO) procedure for QAP which performs well on the QAPLIB instances [10]. We developed a

© Springer International Publishing Switzerland 2016
M.J. Blesa et al. (Eds.): HM 2016, LNCS 9668, pp. 47–61, 2016.
DOI: 10.1007/978-3-319-39636-1_4

cooperative parallel version of this method, a process which was eased thanks to our Cooperative Parallel Local Search (CPLS) framework [11,12] for which we developed an implementation in the X10 programming language [13,14]. This solver (called ParEO) behaves very well on the set of 33 hardest and largest instances of QAPLIB. Using 128 cores and within a short time limit of 5 min, ParEO is able to find the best known solution (BKS) in each replication for 15 problems. Only for 8 instances is the BKS never reached. Recent research shows that the most promising way to improve QAP resolution is to resort to hybrid procedures, in order to benefit from the strengths of different classes of heuristics. Such is the case of hybrid genetic algorithms (a.k.a memetic algorithms) [7]. The price to pay for this improvement is a significant increase in the complexity of the resulting solver code. In any case, many of the best known existing methods can be easily parallelized thanks to our CPLS framework.

In this paper we propose an alternative approach for hybridization: we resort to cooperation and parallelism to get "the best of both worlds". To this end, the parallel instances of different heuristics communicate their best solutions during execution, and are able to forgo the current computation and *adopt* a better solution (hoping it will converge faster). The expected behavior is that a solution which appears to be stagnating inside one solver can be improved by another heuristic. When the second solver can no longer improve on this (imported) solution, maybe the original one can, once again, improve the solution yet a bit more, and so on. It is worth noticing that when the first solver sends its current solution, it continues to work on it until it adopts an external solution, itself. This *cooperative portfolio* approach behaves like a hybrid solver while retaining the original simplicity of each solver. This is particularly true inside the CPLS framework since solvers need not be aware about the (nature of) other solvers.

We implemented such a hybrid solver on top of the X10 version of CPLS, combining two different solvers: Taillard's robust tabu search (RoTS) [15] and our EO-QAP [10] method. We have chosen these two solvers because they are simple and also because it turned out that they present complementary strengths: roughly speaking RoTS is stronger in intensifying the search in a given region while EO-QAP is better at widely diversifying the search. The resulting hybrid cooperative solver (called ParEOTS) displays very good performance, as we shall see further. We show that it scales very well, exhibiting a linear speedup when increasing the number of cores. This solver behaves much better than the cooperative versions of both EO-QAP and RoTS alone. On the 33 hardest instances of QAPLIB, using 128 cores and a time limit of 5 min, ParEOTS is able to find the BKS for 26 problems at each replication. Even for the 7 other problems, the quality of returned solutions (measured as a percentage of average solution over the BKS), is significantly improved. We also test ParEOTS on Palubeckis' InstXX instances and on Drezners dreXX instances. Moreover we provide optimal solutions for several InstXX instances and for dre90, dre100 and dre132.

The rest of the paper is organized as follows: Sect. 2 discusses QAP, RoTS and EO-QAP. Section 3 presents our parallel hybrid solver. Several experimental results are laid out and discussed in Sect. 4 and we conclude in Sect. 5.

2   Background

In this section we recall some background topics: the Quadratic Assignment Problem (QAP) and the two heuristics we plan to combine: RoTS and EO-QAP.

2.1   QAP

Since its introduction in 1957, QAP has been widely studied and several surveys are available [2,3,16,17]. A QAP problem of size n consists of two $n \times n$ matrices (a_{ij}) and (b_{ij}). Solving such a problem consists in finding a permutation π of $\{1, 2, \ldots n\}$, minimizing the objective function: $F(\pi) = \sum_{i=1}^{n} \sum_{j=1}^{n} a_{ij} \cdot b_{\pi_i \pi_j}$. In facility location problems, the a matrix represents inter-facility flows and b encodes the inter-location distances. Moreover, QAP can be also used to model scheduling, chip placement and wiring on a circuit board, to design typewriter keyboards, for process communications, for turbine runner balancing among many other applications [2,18].

QAP is computationally very difficult: it is a discrete problem, the objective function contains products of variables and the theoretical search space of an instance of size n has a size $n!$. QAP has been proved to be NP-hard [19] (the traveling salesman problem can be formulated as a QAP) and there is no ϵ-approximation algorithm for QAP (unless $P = NP$). In practice, this means that QAP is one of the toughest combinatorial optimization problems, and one with several real-life applications.

QAP can be (optimally) solved with exact methods like dynamic programming, cutting plane techniques and branch & bound algorithms for medium sizes, e.g. $n \leq 30$. For larger problems, (meta)heuristics are the most efficient tool. Over the last decades several metaheuristics were successfully applied to QAP: tabu search, simulated annealing, genetic algorithms, GRASP, ant-colonies [20].

2.2   RoTS: A Tabu Search Procedure for QAP

Tabu search as proposed by Glover [21] has been widely used since the 1990 to tackle QAP. Unquestionably, one of the most important algorithms for QAP is Taillard's robust tabu search [15] (RoTS). This algorithm uses an adaptive short-term memory for the tabu list by recording the value assigned to a variable for a while (in order to prevent "reverse assignments"). It also uses a clever *aspiration criterion* (needed to authorize a tabu move to be performed in special circumstances, e.g. if it improves on the best solution found so far). RoTS also incorporates a long-term memory to ensure a form of diversification, by encouraging moves towards to not yet visited regions. RoTS only requires two user parameters to be tuned: the *tabu tenure factor* (controlling the time an element remains tabu) and the *aspiration factor* both of which influence the adaptive memory ([22] provides good references values for these parameters). In practice RoTS is tremendously effective on a wide variety of QAP instances, being able to quickly find high quality solutions. Several BKS for QAPLIB instances have been discovered and/or improved by RoTS. A key feature explaining its speed is that the cost of a solution resulting from a swap can be computed incrementally and further optimized using a tabling mechanism. This results in an evaluation in $O(n^2)$, while the naïve algorithm is in $O(n^3)$. In addition, Taillard put the source code in the public domain. All these reasons explain the fact that RoTS is directly or indirectly at the root of many other methods to solve QAP [4,23].

2.3   EO-QAP: An Extremal Optimization Procedure for QAP

Extremal Optimization (EO) is a metaheuristics inspired by self-organizing processes often found in nature [24–26]. EO is based on the concept of *Self-Organized Criticality* (SOC) initially proposed by Bak and on the Bak-Sneppen's model [27]. In this model of biological evolution, *species* have a *fitness* $\in [0,1]$ (0 representing the worst degree of adaptation). At each iteration, the species with the worst fitness is eliminated (or forced to mutate). This change affects its fitness but also the fitness of all other species connected to this "culprit" element. This results in an *extremal* process which progressively eliminates the least fit species (or forces them to mutate). Repeating this process eventually leads to a state where all species have a good fitness value, i.e. a SOC. EO follows this line: it inspects the current *configuration* (assignment of variables), selects one of the *worst* variables (according to their fitness) to *mutate*. For this, it ranks the variables in increasing order of fitness (the worst variable has thus a rank $k = 1$) and then resorts to a *Probability Distribution Function* (PDF) over the ranks k to chose the culprit element. This PDF introduces uncertainty in the search process. The original EO proposes a power-law: $P(k) = k^{-\tau}(1 \le k \le n)$. This PDF takes a single parameter τ which is problem-dependent. Depending on the value of τ, EO provides different search strategies from pure random walk ($\tau = 0$) to deterministic (greedy) search ($\tau \to \infty$). With an adequate value for τ, EO cannot be trapped in local minima since any variable is susceptible to mutate (even if the worst are privileged). This parameter can be tuned by the user (a default value is $\tau = 1 + \frac{1}{ln(n)}$).

EO displays several a priori advantages: it is a simple metaheuristic, it is controlled by only one free parameter (a fine tuning of several parameters becomes quickly tedious) and it does not need to be aware about local minima. Nevertheless, EO has been successfully applied to large-scale optimization problems like graph bi-partitioning, graph coloring or the traveling salesman problem [25].

Recently, we proposed EO-QAP: an EO procedure for QAP [10]. One notable extension we brought to the original EO is to propose different PDFs and to allow the user to chose the most adequate one for a given problem. The sequential procedure performs well on the whole set of QAPLIB instances: 68 instances are *solved* (i.e. the BKS could be reached) at each execution, 41 are solved at least once and 25 never. The independent parallel version improves the situation significantly: 33 additional instances are systematically solved, 14 are partially solved and 19 remain unsolved. To tackle this remaining set of 33 instances (14+19) we experimented with cooperative parallelism (this version is called ParEO). In the same time limit, ParEO is able to systematically solve 15 new instances and 18 are solved at least once (8 remain unsolved).

3   A Cooperative Parallel Hybrid Method

We propose an alternative approach for constructing hybrid search methods, resorting on our Cooperative Parallel Local Search Framework (CPLS) [11, 12], to provide the hybridization. In a nutshell, the procedure amounts to having several workers, each following its own strategy, some of which are significantly different from others. The cooperative framework oversees every worker, and makes it possible for it to contribute and benefit from the global effort, by managing a pool of best solution candidates (the *elite pool*). The fact that the framework is parallel entitles it to obtain performance benefits by just increasing the count of compute units (cores.) Moreover, the workers themselves need to have little or no knowledge of the environment they are running under.

To test these ideas, we experimented with a solver for QAP – an admittedly difficult problem – for which the individual metaheuristic we chose are our EO-QAP algorithm and the RoTS method.

3.1   Cooperative Parallel Local Search

Parallel local search methods have been proposed in the past [28–30]. Here we focus on *multi-walk* methods (also called *multi-start*) which consist in a concurrent exploration of the search space, either *independently* or *cooperatively*, the latter being achieved with communication between processes. The *Independent Multi-Walks* method (IW) [31] is easiest to implement since the solver instances need not communicate with each other. However, the resulting gain tends to flatten when scaling beyond about a hundred processors [32], largely because the inherent diversity which brings about the speedups is not sufficient. In the *Cooperative Multi-Walks* (CW) method [33], the solver instances exchange information (through communication), hoping to hasten the search process. However, the design and implementation of an efficient such method is a very challenging task: choices abound concerning the communication which impact each other, many of which are problem-dependent [33].

We designed the Cooperative Parallel Local Search (CPLS) framework [11, 12]. This framework, available as an open source library in the X10 programming language, allows the programmer to tune the search process through an extensive set of parameters which, at present, statically condition the execution. CPLS augments the IW strategy with a tunable communication mechanism, which allows for the cooperation between the multiple solver instances to seek either an intensification or diversification strategy in the search. At present, the tuning process is done manually: we have not yet experimented with parameter self-adaptation in the CPLS framework (still an experimental feature).

The basic component of CPLS is the *explorer node* which consists in a local search-based solver instance. The point is to use all the available processing units by mapping each *explorer node* to a physical core. Explorer nodes are grouped into *teams*, of size NPT (see Fig. 1). This parameter is directly related to the trade-off between intensification and diversification. NPT can take values from 1 to the maximum number of cores. When NPT is equal to 1, the framework coincides with the IW strategy, it is expected that each 1-node team be working on a different region of the search space, without any effort to seek parallel intensification. When NPT is equal to the maximum number of nodes (creating only 1 team in the execution), the framework is mainly geared towards parallel intensification (however a certain amount of diversification is inherently provided by parallelism, between 2 cooperation actions).

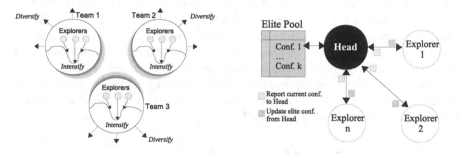

Fig. 1. CPLS framework structure

Each team seeks to *intensify* the search in the most promising neighborhood found by any of its members. The parameters which guide the intensification are the *Report Interval* (R) and *Update Interval* (U): every R iterations, each explorer node sends its current configuration and the associated cost to its *head node* (*report event*). The head node is the team member which collects and processes this information, retaining the best configurations in an *Elite Pool* (EP) whose size $|EP|$ is parametric. Every U iterations, explorer nodes randomly retrieve a configuration from the EP, in the head node (*update event*). An explorer node may *adopt* the configuration from the EP, if it is "better" than its own current configuration, with a probability *pAdopt*. Simultaneously, the teams implement a mechanism to cooperatively *diversify* the search, i.e. they try to extend the search to different regions of the search space.

Typically, each problem benefits from intensification and diversification to some extent. Therefore, the tuning process of the CPLS parameters seeks to provide an appropriate balance between the use of the intensification and diversification mechanisms, in hope of reaching better performance than the non-cooperative parallel solvers (i.e. independent multi-walks). A detailed description of this framework may be found in [11].

3.2 Using the CPLS Framework for Hybridization

The current X10 implementation of the CPLS framework already supports the use of multiple metaheuristics. Adding a new one is simple because CPLS provides useful abstraction layers and handles communication. Adding a new metaheuristic comes down to slightly adapt the sequential algorithm: every R iterations it has to send its current configuration to the Elite Pool and, every U

iterations, it needs to retrieve a configuration from the pool, which it may subsequently adopt (with probability *pAdopt*), should it be better than the current one. The overall resulting solver is thus composed of several instances of the *same* metaheuristic running in parallel, which cooperate by communicating in order to faster converge to a solution. To date, CPLS includes cooperative parallel versions of three different methods: *Adaptive Search*, *Extremal Optimization* and *Tabu Search*. In the present work, we go one step beyond and propose a new usage of the CPLS framework in order to obtain an hybrid parallel solver. For this, individual workers run instances of *different* metaheuristics, while still collaborating by communicating with the head node. The basic idea of running different metaheuristics in parallel exchanging elite solutions has been proposed [28,34] but only from a general and theoretical point of view. This can also be viewed as a *portfolio* approach [35] augmented with cooperation.

We chose to experiment with this form of hybridization on QAP combining two metaheuristics: our EO-QAP procedure and the RoTS method, resulting in a solver we call ParEOTS. The communication strategies of CPLS remain unchanged, ensuring cooperation between the explorers which now happen to be running different methods. Figure 2 presents possible interactions due to cooperation and the implementation of the hybrid strategy. The team's EP will now contain configurations stemming from explorers running different heuristics.

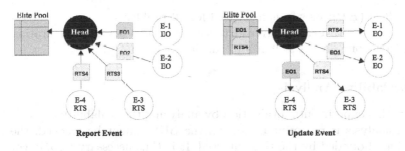

Fig. 2. Hybridization in CPLS : combining EO-QAP and RoTS

Here is a possible scenario: inside the same team, an instance E_1 of EO-QAP reports a good configuration C_1 to the EP. Later, an instance R_1 of RoTS retrieves C_1, improves on it (RoTS being strong at intensification) and obtains a better configuration C_2, on which it reports back to the EP. Later, C_2 gets adopted by an instance E_2 which, being in a diversification phase, moves to a faraway search region, which may provide yet better solutions. Obviously, other scenarios are possible, e.g. when another EO-QAP explorer E_3 also retrieves C_1 (provided by EO) it gives a "second chance" to this configuration (due to its internal stochastic state it can further improve this configuration). The whole system behaves as a hybrid solver, benefiting from cross-fertilization due to the inherent diversity of the search strategies.

4   Experimental Evaluation

In this section we present an experimental evaluation of our hybrid parallel method (source code, instances and new solutions will be soon available from http://cri-hpc1.univ-paris1.fr/qap/). All experiments have been carried out on a cluster of 16 machines, each with 4×16-core AMD Opteron 6376 CPUs running at 2.3 GHz and 128 GB of RAM. The nodes are interconnected with InfiniBand FDR $4\times$ (i.e. 56 GBPS). We had access to 4 nodes and used up to 32 cores per node, i.e. 128 cores. Each problem is executed 10 times stopping as soon as the BKS (which is sometimes the optimum) is found. This execution is done with a short time limit of 5 min (in case the BKS is not reached). Such experiments give an interesting information about the quality of solutions quickly obtainable. All times are given either in seconds for small values (as a decimal number) or in a human readable form as mm:ss or hh:mm:ss). The relevant CPLS parameters controlling the cooperation are (as per [11]):

- *Team Size (NPT)*: we fixed it to $NPT = 16$. There are thus 8 teams composed of 16 explorer nodes ; 8 running a EO-QAP solver and 8 running RoTS solver. This is constant over all problems. We did not yet experiment with other splits.
- *Report and Update Interval (R and U)*: we manually tuned U and usually fix $R = U/2$.
- *Elite Pool (EP)*: its size is fixed to 4 for all problems.
- *pAdopt*: is set to 1. Any solver instance receiving a better configuration than its current one always switches to the new one.

4.1   Scalability Analysis

We start this experimental evaluation by analyzing the scalability of ParEOTS. Such an analysis is not easy, because if the BKS cannot be reached, the runtime is only bounded by the timeout used. It is thus necessary to only consider problems that can be systematically solved by the EO sequential solver (to have a reference time using 1 core). We selected two instances of QAPLIB which require the longest sequential time: `tai35a` solved on average in 42.399 s and `lipa70a` solved in 57.737 s. We then ran these problems with ParEOTS, varying the number of cores from 2 to 128. Figure 3 presents the speedup data and curves obtained with our algorithm (using a log-log scale). The *Ideal* curve corresponds to linear speedup: time is halved when the number of cores is doubled. For both problems the speedup is linear. Using 128 cores, the best speedup is 126, obtained for `tai35a` whose execution time now only requires 0.336 s.

4.2   Evaluation on QAPLIB

We here evaluate the performance of our hybrid solver ParEOTS on a set of 33 hard instances of QAPLIB. We selected this set because it is the most difficult set for the independent parallel version of our EO procedure [10]. In addition to raw

Cores	tai35a		lipa70a	
	time	speedup	time	speedup
1	0:42	1.0	0:57	1.0
2	0:33	1.3	0:18	3.2
4	0:20	2.1	0:17	3.4
8	8.9	4.8	8.4	6.9
16	6.3	6.8	3.8	15.2
32	2.6	16.6	2.1	27.5
64	1.4	31.4	1.1	54.3
128	0.3	126.0	0.5	106.3

Fig. 3. Speedup profile using the Hybrid CPLS on two QAPLIB instances

performance, and for validation purpose, we also want to assess the gain obtained with the hybrid version compared the cooperative parallel versions of its two components: ParEO and ParRoTS (also written in X10 within CPLS). For this, all 3 systems are run under the same conditions (see Sect. 4). Obviously, ParEO runs 128 instances of our EO procedure, ParRoTS runs 128 instances of RoTS while ParEOTS executes 64 instances of EO-QAP and 64 of RoTS. To measure the *hybrid performance* we focus on the number of BKS found by each parallel solver. When running 50 % of EO and 50 % of RoTS we define a *low threshold* (*low*) as the average of #BKS found by both parallel solvers. This corresponds to what can be normally expected. Below this value, the hybrid solver is ineffective. Above, it already performs well. Moreover, we define a *high threshold* (*high*) as the maximum of the #BKS of both solvers. Above this value, the hybrid solver performs at least as well as the best single solver (a hybrid solver without gain would need twice the number of cores to obtain such a performance). Obviously *low* and *high* can be generalized to an hybridization involving more than 2 solvers. For a given problem, executed n times, the performance (*hperf*) of the hybrid solver reaching #*bks* times the BKS is defined as follows:

$$
hperf = \begin{cases} \frac{\#bks-low}{low}, & \text{if } \#bks < low \\ \frac{\#bks-low}{high-low}, & \text{if } low < \#bks < high \text{ and } low \neq high \\ 1+\frac{\#bks-high}{n-high}, & \text{if } high \leq \#bks \text{ and } n \neq high \\ 1, & \text{if } high = \#bks = n \end{cases} \tag{1}
$$

The performance ranges in $[-1, 2]$. if $hperf < 0$ the hybrid solver is ineffective on that problem. For $hperf \in [0, 1)$ the performance is acceptable and when $hperf \in [1, 2]$ the performance is very good.

low	high	very high
-1	0	1	2

Table 1. ParEOTS on QAPLIB and comparison with ParEO and ParRoTS

	ParEOTS			hperf	ParEO				ParRoTS			
	$\#_{bks}$	APD	time	$\#_{ad.}$		$\#_{bks}$	APD	time	$\#_{ad.}$	$\#_{bks}$	APD	time	$\#_{ad.}$
els19	**10**	0.000	0.0	2.6	1.00	**10**	0.000	0.0	0.2	**10**	0.000	0.0	0.3
kra30a	**10**	0.000	0.0	3.9	1.00	**10**	0.000	0.0	2.6	**10**	0.000	0.0	3.6
sko56	**10**	0.000	1.5	0.3	1.00	**10**	0.000	4.8	2.5	**10**	0.000	0.6	0.0
sko64	**10**	0.000	1.7	0.3	1.00	**10**	0.000	4.8	1.5	**10**	0.000	1.3	0.0
sko72	**10**	0.000	8.7	1.0	1.00	**10**	0.000	0:13	1.4	**10**	0.000	0:16	1.7
sko81	**10**	0.000	0:24	1.8	1.00	7	0.008	1:58	9.4	**10**	0.000	1:06	4.6
sko90	**10**	0.000	1:32	4.8	1.00	**10**	0.000	1:32	5.0	7	0.002	1:54	5.3
sko100a	**10**	0.000	1:09	1.3	2.00	5	0.012	3:44	4.2	7	0.002	2:46	3.3
sko100b	**10**	0.000	0:45	0.8	1.00	8	0.001	2:26	2.6	**10**	0.000	1:02	0.6
sko100c	**10**	0.000	0:56	1.0	1.00	**10**	0.000	2:25	2.4	6	0.001	3:12	3.6
sko100d	**10**	0.000	1:03	1.1	1.00	6	0.014	3:20	3.6	**10**	0.000	0:37	0.2
sko100e	**10**	0.000	0:47	0.9	1.00	**10**	0.000	1:43	1.6	5	0.002	2:47	3.0
sko100f	**10**	0.000	0:57	0.9	2.00	4	0.011	4:05	4.8	5	0.003	3:42	4.3
tai40a	**10**	0.000	1:26	1.6	1.00	7	0.022	2:51	3.4	**10**	0.000	1:04	1.0
tai50a	3	0.077	4:24	3.5	−0.33	5	0.026	3:28	2.4	4	0.044	4:11	2.5
tai60a	3	0.146	4:15	0.9	1.13	2	0.132	4:45	1.9	0	0.297	5:00	2.0
tai80a	0	0.364	5:00	4.9	1.00	0	0.385	5:00	1.0	0	0.605	5:00	1.0
tai100a	0	0.298	5:00	2.0	1.00	0	0.297	5:00	3.0	0	0.567	5:00	5.0
tai20b	**10**	0.000	0.0	1.0	1.00	**10**	0.000	0.0	0.8	**10**	0.000	0.0	0.3
tai25b	**10**	0.000	0.0	0.5	1.00	**10**	0.000	0.6	17.0	**10**	0.000	0.0	0.0
tai30b	**10**	0.000	0.1	1.9	1.00	**10**	0.000	0.1	3.0	**10**	0.000	0.1	1.2
tai35b	**10**	0.000	0.3	4.3	1.00	**10**	0.000	0.7	14.2	**10**	0.000	0.2	1.9
tai40b	**10**	0.000	0.1	0.6	1.00	**10**	0.000	0.1	0.4	**10**	0.000	0.2	2.0
tai50b	**10**	0.000	2.6	1.2	1.00	2	0.214	4:26	4.5	**10**	0.000	2.1	0.0
tai60b	**10**	0.000	4.6	1.2	1.00	3	0.205	4:16	2.6	**10**	0.000	5.3	0.0
tai80b	**10**	0.000	0:53	1.6	2.00	0	1.192	5:00	8.8	5	0.002	3:06	6.0
tai100b	**10**	0.000	1:11	0.7	2.00	0	0.465	5:00	5.5	2	0.035	4:10	4.8
tai150b	0	0.061	5:00	0.7	1.00	0	1.088	5:00	1.5	0	0.103	5:00	0.3
tai64c	**10**	0.000	0.0	0.0	1.00	**10**	0.000	0.0	0.3	**10**	0.000	0.0	0.0
tai256c	0	0.178	5:00	2.2	1.00	0	0.263	5:00	1.3	0	0.266	5:00	1.5
tho40	**10**	0.000	0.5	0.0	1.00	**10**	0.000	1.2	0.2	**10**	0.000	0.4	0.0
tho150	1	0.007	4:51	2.0	1.10	0	0.144	5:00	1.7	0	0.019	5:00	1.9
wil100	**10**	0.000	1:37	1.9	2.00	0	0.061	5:00	5.4	6	0.001	2:16	2.4
Summary	**267**	**0.034**	**1:24**	**1.6**	**1.12**	199	0.138	2:28	3.7	227	0.059	1:53	1.9

Table 1 presents the results. The parameters used for EO are the same as in [10]. For RoTS we generally use a tabu tenure $= 8n$ and an aspiration $= 4n^2$. The table reports, for each solver, the number of times out of 10 runs the BKS was reached ($\#bks$), the Average Percentage Deviation (APD) which is relative deviation percentage computed as follows: $100 \times \frac{Avg - BKS}{BKS}$ (where Avg is the average of the 10 found costs), the average execution time (average of the 10 wall times for one instance) and the numbers of adoptions done by the winning explorer (#ad.). The performance value is also reported. The last row presents the averages of each column (or sums for #*bks* columns).

It is worth noticing that the overall performance of the cooperative parallel version of the 2 base solvers using a short time limit is rather good. Even so, the hybrid solver clearly outperforms them. Focusing on #BKS, it provides high performance (*hperf* $\geq$ 1) for 32 instances (only for tai50a does it behave worse than its two components). Moreover, in 4 cases it obtains a *hperf* = 2 corresponding to cases where it performs much better than both individual solvers (to such an extent that it obtains the perfect score #BKS = 10). It found the BKS at each replication for 26 problems; this is much better than ParEO (15) and ParRoTS (18). In only 4 cases, could ParEOTS not reach the BKS: this number is 8 for ParEO and 6 for ParRoTS. It is worth noticing that even in these 4 cases, the hybridization is still effective since the APD is lower than for its components. For instance, on the very difficult problem tai256c, the hybridization cannot solve the problem but the APD is 0.178 while it is around 0.263 for both components. Another remarkable case is tho150, for which the hybridization is very effective. The average APD is now 0.007 (0.144 for ParEO and 0.019 for ParRoTS). In fact, it turns out that this problem could even be solved once.

The "summary" row reports interesting numbers. All in one, the average APD of ParEOTS is 0.034 which is much better than 0.138 for ParEO and 0.059 for ParRoTS. Regarding execution times, it is a good surprise to see that the increase of quality does not hamper the speed. In fact, with an average execution time of 85 s the hybrid solver is faster than ParEO (148 s) and ParRoTS (113 s).

4.3 Testing on Palubeckis' Instances

In 2000, Palubeckis proposed a new hard problem generator with known optimum [36] and provided a set of 10 hard instances called Inst*XX*. Few results have been published about experiments with them. Palubeckis reports the best solutions found by a repeated local search procedure (called multi-start descent or MSD). In [37] the authors propose an Ant Colony Optimization algorithm (QAP-ACO) and test it on these instances (in this work these instances are called palu*XX*).

We experimented in the same setting as previously: with 128 cores and a time limit of 5 min. Table 2 displays the results for 3 solvers. In addition to the APD we also provide the *best* cost value found among the 10 runs. Data is taken from the aforementioned articles. We also provide execution times for QAP-ACO for indicative purposes.

Even with a limit of 5 min, the performances of ParEOTS are very good. The optimum is reached for problems whose size $n \leq 100$. In addition, for all $n \leq 80$ ParEOTS reaches the optimum at each replication. For sizes $n > 100$, clearly a limit of 5 min is too short to reach the optimum. Nevertheless, the obtained solutions are of good quality with an APD around 1.12 %: 2–3 times better than challengers. It is worth noticing that for $n > 20$ all published best obtained

Table 2. ParEOTS on Palubeckis' instances (128 cores, timeout 5 m)

	opt.	ParEOTS				QAP-ACO				MSD		
		$\#_{bks}$	APD	best value	time	$\#_{bks}$	APD	best value	time	$\#_{bks}$	APD	best value
Inst20	81536	10	0.000	81536	0.0	0	0.340	81817	1.2	10	0.000	81536
Inst30	271092	10	0.000	271092	0.1	0	0.580	272654	0:10	0	0.364	272080
Inst40	837900	10	0.000	837900	4.0	0	0.360	840930	1:02	0	0.287	840308
Inst50	1840356	10	0.000	1840356	0:17	0	0.380	1847422	3:46	0	0.354	1846876
Inst60	2967464	10	0.000	2967464	1:07	0	0.390	2978898	10:05	0	0.362	2978216
Inst70	5815290	10	0.000	5815290	2:07	0	0.300	5832460	24:24	0	0.287	5831954
Inst80	6597966	10	0.000	6597966	1:56	0	0.310	6618736	50:42	0	0.308	6618290
Inst100	15008994	1	0.120	15008994	5:00	0	0.270	15048806	1:41:02	0	0.256	15047406
Inst150	58352664	0	0.126	58414888	5:00					0	0.198	58468204
Inst200	75405684	0	0.125	75498892	5:00					0	0.183	75543960

solutions are improved (in bold font in the table). Regarding execution times, ParEOTS also outperforms its competitors.

4.4   Testing on Drezner's Instances

In 2005, Drezner and al. designed new QAP instances with known optimum but specifically ill conditioned to be difficult for metaheuristic methods [38]. The authors reports the best solutions found by a powerful compounded hybrid genetic algorithm (called CHG in what follows). The instances are really difficult and only very recently were some results published by Acan and Ünveren with a *great deluge* algorithm (called TMSGD) [39]. These hard instances are thus an interesting challenge for our hybrid solver.

We ran it under the same conditions as before: using 128 cores and with a time limit of 5 s. Table 3 presents the results for 3 solvers. Data is taken from

Table 3. ParEOTS on Drezner's instances (128 cores, timeout 5 m)

	opt.	ParEOTS				TMSGD				CHG		
		$\#_{bks}$	APD	best	time	$\#_{bks}$	APD	best	time	$\#_{bks}$	APD	best
dre15	306	10	0.000	306	0.0	10	0.000	306	2.1			
dre18	332	10	0.000	332	0.0	10	0.000	332	7.4			
dre21	356	10	0.000	356	0.0	10	0.000	356	0:18			
dre24	396	10	0.000	396	0.0	10	0.000	396	0:56			
dre28	476	10	0.000	476	0.1	10	0.000	476	1:18			
dre30	508	10	0.000	508	0.1	10	0.000	508	2:36	10	0.00	508
dre42	764	10	0.000	764	0.692	6	0.25	764	8:51	9	1.34	764
dre56	1086	10	0.000	1086	5.6	3	3.556	1086	18:39	3	17.46	1086
dre72	1452	10	0.000	1452	0:26	0	8.388	1512	47:06	1	27.28	1452
dre90	1838	9	0.968	1838	2:47	0	10.979	1959	1:36:33	0	33.88	2218
dre110	2264	6	6.334	2264	3:43	0	15.123	2479	2:41:25			
dre132	2744	1	22.784	2744	4:54	0	17.553	3023	3:31:07			

the above mentioned articles (in the case of CHG each problem was executed 20 times, presented #BKS are divided by 2 for normalization). We also provide execution times for TMSGD for indicative purposes (TMSGD was run on a 2.1 GHz PC).

The performance of ParEOTS is very good: all problems could be optimally solved, and, to the best of our knowledge, this is the first time that an optimal solution is found for dre90, dre110 and dre132. TMSGD performs better than CHG (but the CHG experiment is old). Regarding execution times, ParEOTS needs 2:47 to solve dre90 while TMSGD cannot solve it even using 1:36:33 (CHG reports one hour for dre90 and also fails to find the optimum).

5 Conclusion and Further Work

We set out to construct a hybridized solver by resorting to a parallel and cooperative multi-walk scheme, which relies on the CPLS framework to provide both the cooperation and the parallel or distributed execution.

As a testbed for the idea, we chose to tackle the Quadratic Assignment Problem, because it is recognized as a very difficult problem of significant practical interest and also because benchmark instances abound in the literature. For this we designed ParEOTS: a hybrid cooperative parallel solver combining two methods: our Extremal Optimization algorithm and Taillard's robust tabu search. This hybrid solver is much more efficient than any of its two individual base solvers. Regarding QAPLIB, our hybrid solver is able to reach the best known solution (BKS) for all instances except 4. In most cases it is even able to systematically find the BKS at each replication. Even then, for the 4 not fully solved hardest instances (tai80a, tai100a, tai150b and tai256c), the solutions obtained are very close to the BKS. We also tested the solver on other hard instances. The results on Palubeckis' instances are very good: for the first time, ParEOTS optimally solved all instances up to a size $n = 100$ (prior to this work only optimal solutions for $n = 20$ were known). We discovered optimal solutions for sizes $n = 30..100$ and 2 new best obtained solutions for $n = 150$ and $n = 200$. Regarding Drezner's instances, the results are even better: we discovered optimal solutions for all instances (including dre90, dre110 and dre132). This is the first time that optimal solutions for these 3 instances are published.

From our experiments, it became clear that: (1) the coding effort for building a hybrid solver is much lower with our approach than for existing hybrid algorithms, and (2) the performance gain over competing approaches is very significant. The latter aspect can be construed as a sort of evolutionary algorithm, one which blends phenotypes rather than genotypes, all under the supervision of the cooperative framework. As to the former, the changes needed to fit the CPLS scheme are minimal and very simple.

We plan to further explore portfolio approaches, combining more than two types of solver as well as experimenting with techniques for parameter autotuning. Another line entails the induction of solver multiplicity by presenting several instances of the same solver, but set up with different parameters.

Acknowledgments. The authors wish to thank Prof. E. Taillard for providing the RoTS source code and explanations. The experimentation used the cluster of the University of Évora, which was partly funded by grants ALENT-07-0262-FEDER-001872 and ALENT-07-0262-FEDER-001876.

References

1. Koopmans, T.C., Beckmann, M.: Assignment problems and the location of economic activities. Econometrica **25**(1), 53–76 (1957)
2. Commander, C.W.: A survey of the quadratic assignment problem, with applications. Morehead Electron. J. Appl. Math. **4**, 1–15 (2005). MATH-2005-01
3. Bhati, R.K., Rasool, A.: Quadratic assignment problem and its relevance to the real world: a survey. Int. J. Comput. Appl. **96**(9), 42–47 (2014)
4. James, T., Rego, C., Glover, F.: Multistart tabu search and diversification strategies for the quadratic assignment problem. IEEE Trans. Syst. Man Cybern. Part A: Syst. Hum. **39**(3), 579–596 (2009)
5. Benlic, U., Hao, J.K.: Breakout local search for the quadratic assignment problem. Appl. Math. Comput. **219**(9), 4800–4815 (2013)
6. Drezner, Z.: The extended concentric tabu for the quadratic assignment problem. Eur. J. Oper. Res. **160**(2), 416–422 (2005)
7. Drezner, Z.: Extensive experiments with hybrid genetic algorithms for the solution of the quadratic assignment problem. Comput. Oper. Res. **35**(3), 717–736 (2008)
8. James, T., Rego, C., Glover, F.: A cooperative parallel tabu search algorithm for the quadratic assignment problem. Eur. J. Oper. Res. **195**, 810–826 (2009)
9. Tosun, U.: On the performance of parallel hybrid algorithms for the solution of the quadratic assignment problem. Eng. Appl. Artif. Intell. **39**, 267–278 (2015)
10. Munera, D., Diaz, D., Abreu, S.: Solving the quadratic assignment problem with cooperative parallel extremal optimization. In: Chicano, F., et al. (eds.) EvoCOP 2016. LNCS, vol. 9595, pp. 251–266. Springer, Heidelberg (2016). doi:10.1007/978-3-319-30698-8_17
11. Munera, D., Diaz, D., Abreu, S., Codognet, P.: A parametric framework for cooperative parallel local search. In: Blum, C., Ochoa, G. (eds.) EvoCOP 2014. LNCS, vol. 8600, pp. 13–24. Springer, Heidelberg (2014)
12. Munera, D., Diaz, D., Abreu, S., Codognet, P.: Flexible cooperation in parallel local search. In: Symposium on Applied Computing (SAC), pp. 1360–1361. ACM Press, New York (2014)
13. Charles, P., Grothoff, C., Saraswat, V., Donawa, C., Kielstra, A., Ebcioglu, K., Von Praun, C., Sarkar, V.: X10: an object-oriented approach to non-uniform cluster computing. In: SIGPLAN Conference on Object-oriented Programming, Systems, Languages, and Applications, pp. 519–538. ACM, San Diego (2005)
14. Saraswat, V., Tardieu, O., Grove, D., Cunningham, D., Takeuchi, M., Herta, B.: A Brief Introduction to X10 (for the High Performance Programmer). Technical report (2012)
15. Taillard, É.D.: Robust taboo search for the quadratic assignment problem. Parallel Comput. **17**(4–5), 443–455 (1991)
16. Burkard, R.E.: Quadratic assignment problems. In: Pardalos, P.M., Du, D.Z., Graham, R.L. (eds.) Handbook of Combinatorial Optimization, 2nd edn, pp. 2741–2814. Springer, New York (2013)
17. Loiola, E.M., de Abreu, N.M.M., Netto, P.O.B., Hahn, P., Querido, T.M.: A survey for the quadratic assignment problem. Eur. J. Oper. Res. **176**(2), 657–690 (2007)

18. Zaied, A.N.H., Shawky, L.A.E.F.: A survey of quadratic assignment problems. Int. J. Comput. Appl. **101**(6), 28–36 (2014)
19. Sahni, S., Gonzalez, T.: P-complete approximation problems. J. ACM **23**(3), 555–565 (1976)
20. Said, G., Mahmoud, A.M., El-Horbaty, E.S.M.: A comparative study of meta-heuristic algorithms for solving quadratic assignment problem. Int. J. Adv. Comput. Sci. Appl. (IJACSA) **5**(1), 1–6 (2014)
21. Glover, F., Laguna, M.: Tabu Search. Kluwer Academic Publishers, Boston (1997)
22. Taillard, É.D.: Comparison of iterative searches for the quadratic assignment problem. Location Sci. **3**(2), 87–105 (1995)
23. Misevicius, A.: A tabu search algorithm for the quadratic assignment problem. Comput. Optim. Appl. **30**(1), 95–111 (2005)
24. Boettcher, S., Percus, A.: Nature's way of optimizing. Artif. Intell. **119**(1–2), 275–286 (2000)
25. Boettcher, S., Percus, A.G.: Extremal optimization: an evolutionary local-search algorithm. In: Bhargava, H.K., Ye, N. (eds.) Computational Modeling and Problem Solving in the Networked World, vol. 21. Springer, New York (2003)
26. Boettcher, S.: Extremal Optimization. In: Hartmann, A.K., Rieger, H. (eds.) New Optimization Algorithms to Physics, pp. 227–251. Wiley-VCH Verlag, Berlin (2004)
27. Bak, P., Sneppen, K.: Punctuated equilibrium and criticality in a simple model of evolution. Phy. Rev. Lett. **71**(24), 4083–4086 (1993)
28. Alba, E.: Parallel Metaheuristics: A New Class of Algorithms. Wiley-Interscience, New York (2005)
29. Alba, E., Luque, G., Nesmachnow, S.: Parallel metaheuristics: recent advances and new trends. Int. Trans. Oper. Res. **20**(1), 1–48 (2013)
30. Diaz, D., Abreu, S., Codognet, P.: Parallel constraint-based local search on the cell/be multicore architecture. In: Essaaidi, M., Malgeri, M., Badica, C. (eds.) Intelligent Distributed Computing IV. SCI, vol. 315, pp. 265–274. Springer, Heidelberg (2010)
31. Verhoeven, M., Aarts, E.: Parallel local search. J. Heuristics **1**(1), 43–65 (1995)
32. Caniou, Y., Codognet, P., Richoux, F., Diaz, D., Abreu, S.: Large-scale parallelism for constraint-based local search: the costas array case study. Constraints **20**(1), 1–27 (2014)
33. Toulouse, M., Crainic, T., Sansó, B.: Systemic behavior of cooperative search algorithms. Parallel Comput. **30**, 57–79 (2004)
34. Talukdar, S., Baerentzen, L., Gove, A., De Souza, P.: Asynchronous teams: cooperation schemes for autonomous agents. J. Heuristics **4**(4), 295–321
35. Gomes, C.P., Selman, B.: Algorithm portfolios. Artif. Intell. **126**(1–2), 43–62 (2001)
36. Palubeckis, G.: An algorithm for construction of test cases for the quadratic assignment problem. Informatica Lith. Acad. Sci. **11**(3), 281–296 (2000)
37. Wu, K.C., Ting, C.J., Gonzalez, L.C.: An ant colony optimization algorithm for quadratic assignment problem. In: Asia-Pacific Conference on Industrial Engineering and Management Systems (2011)
38. Drezner, Z., Hahn, P., Taillard, É.: Recent advances for the quadratic assignment problem with special emphasis on instances that are difficult for meta-heuristic methods. Ann. Oper. Res. **139**(1), 65–94 (2005)
39. Acan, A., Ünveren, A.: A great deluge and tabu search hybrid with two-stage memory support for quadratic assignment problem. Appl. Soft Comput. **36**(C), 185–203 (2015)

Investigating Edge-Reordering Procedures in a Tabu Search Algorithm for the Capacitated Arc Routing Problem

Wasin Padungwech$^{(\boxtimes)}$, Jonathan Thompson, and Rhyd Lewis

School of Mathematics, Cardiff University, Cardiff, UK
{padungwechw,thompsonjm1,lewisr9}@cardiff.ac.uk

Abstract. This paper presents two ideas to guide a tabu search algorithm for the Capacitated Arc Routing Problem to a promising region of the solution space. Both ideas involve edge-reordering, although they work in different ways. One of them aims to directly tackle deadheading cycles, and the other tries to reorder edges with the aim of extending a scope of solutions that can be reached from a given solution. Experiments were performed on 134 benchmark instances of various sizes, and the two ideas were shown to have an ability to guide the search to good solutions. Possible issues that may arise when implementing these ideas are also discussed.

1 Introduction

The Capacitated Arc Routing Problem (CARP) is a combinatorial optimisation problem that can be defined as follows: Given a graph with one of its vertices called *the depot*, a cost and a demand for each edge, and a vehicle capacity, the objective of the CARP is to find a minimum-cost set of routes (one route for each vehicle) such that (i) each route contains the depot, (ii) all edges with non-zero demands (called *required edges*) are serviced in precisely one of the routes, and (iii) the total demand in each route does not exceed the capacity. The CARP can be used to model and solve various real-life situations such as rubbish collection, street sweeping, and snow ploughing. It was originally introduced and proved to be NP-hard by Golden and Wong [8].

A wide variety of algorithms have been proposed to solve the CARP, possibly as a result of its real-world applicability. Metaheuristics have been popular choices, and include guided local search [1], scatter search [9], variable neighbourhood search [13], ant colony optimisation [14], memetic algorithms [6,15], and tabu search [2,9,10].

Despite a wide variety of proposed algorithms in the literature, there are still some CARP benchmark instances that remain unsolved, especially those with a relatively large number (347 to 375) of required edges. This suggests that more efficient algorithms for the CARP are still to be found. One key idea could be to find a way to explore a space of solutions efficiently. Traditional neighbourhood moves for the CARP such as removing or inserting edges or swapping edges between routes usually affect only a small number of edges and

© Springer International Publishing Switzerland 2016
M.J. Blesa et al. (Eds.): HM 2016, LNCS 9668, pp. 62–74, 2016.
DOI: 10.1007/978-3-319-39636-1_5

leave other edges untouched (apart from perhaps shifting their orders). It could be beneficial to integrate such moves with a method that can extend the scope of solutions that can be reached from a given solution, thereby increasing the connectivity of the solution space. One possible way to achieve that is to allow edges in a route to be reordered when receiving a new edge from another route. This could better accommodate the new edge and lead to an improvement which might have otherwise required several traditional moves.

Note that allowing edges to be reordered can greatly enlarge a neighbourhood of a given solution. In this paper, we therefore present two ideas that can help a search head towards a promising region of the solution space. The first idea is based on an investigation into deadheading edges, i.e. edges that are not serviced by a vehicle but are used to travel from one serviced edge to another. A route usually contains not only serviced edges but also deadheading edges. If deadheading edges form a cycle, such cycle should be removed provided that the route does not get disconnected as a result. This potentially reorders the edges in the route while definitely improving a solution (assuming non-zero edge costs). This could be particularly useful for large instances with a small ratio of capacity to total demand, where vehicles tend to fill the capacity quickly and have to return to the depot early, resulting in a significant amount of deadheading cost.

The second idea is to reconstruct a route with a given set of required edges. This can be achieved by means of a heuristic algorithm for the Rural Postman Problem, a special case of the CARP in which a single vehicle has large enough capacity to service all edges (see, for example, [4]). This idea was also utilised by Brandão and Eglese in [2]. In their paper, the heuristic is applied to routes that are changed by the best neighbourhood move in each iteration. In this paper, by contrast, the heuristic is integrated with each neighbourhood move, so it is also taken into account when finding the best neighbourhood move.

This paper is organised as follows: A formal definition of the CARP is given in Sect. 2. Section 3 explains how deadheading cycles can occur in a route and introduces a procedure for removing such cycles. Section 4 describes a tabu search algorithm with edge-reordering procedures. Performances of the algorithm are presented and discussed in Sect. 5. Finally, Sect. 6 gives a conclusion and suggestions for future work.

2   Problem Definition and Notation

Given an undirected graph $G = (V, E)$ with a vertex set V and an edge set E, a cost $c(e) \in \mathbb{Z}^+$ and a demand $d(e) \in \mathbb{Z}^+ \cup \{0\}$ for each edge $e \in E$, a vehicle capacity $Q \in \mathbb{Z}^+$, and one of the vertices $v_0 \in V$ regarded as the *depot*, the objective of the CARP is to find a minimum-cost set of routes such that

- each route contains the depot,
- each edge with non-zero demand (called a *required edge*) is serviced in precisely one of the routes, and
- the total demand of serviced edges in each route does not exceed the vehicle capacity.

Here, the number of routes is treated as a variable. Note that the orientation of each edge in a route needs to be specified even if the underlying graph is undirected because this can affect the cost of travelling from one required edge to another. An edge $e = \{i, j\}$ can be traversed in two possible ways: from i to j or from j to i, denoted by directed edges, or arcs, (i, j) and (j, i), respectively. A route R can then be represented as a sequence of arcs $(a_1, a_2, \ldots, a_n)$, where n is the number of arcs that are serviced by R and $a_1, a_2, \ldots, a_n$ are the serviced arcs. This is possible as it is clear that a vehicle should travel between serviced arcs via a shortest path in order to minimise the overall cost. This representation of a route is similar to that in [1].

Let $t(a)$ and $h(a)$ denote the tail and the head of an arc a. For example, if $a = (i, j)$, then $t(a) = i$ and $h(a) = j$. The total cost of the route $R = (a_1, a_2, \ldots, a_n)$ is given by

$$\mathcal{C}(R) = d(v_0, t(a_1)) + \sum_{i=1}^{n-1} d(h(a_i), t(a_{i+1})) + d(h(a_n), v_0) + \sum_{i=1}^{n} c(a_i), \quad (1)$$

where $d(u, v)$ is the cost of a shortest path between vertices u and v. The total demand of R is given by

$$\mathcal{D}(R) = \sum_{i=1}^{n} d(a_i). \quad (2)$$

Using the above notations, the CARP can be presented more formally. Let $x(e, R)$ be a binary variable such that $x(e, R) = 1$ if an edge e is serviced in a route R, and $x(e, R) = 0$ otherwise. Let E_R be the set of required edges. The objective of the CARP is to find a set of routes S that minimises

$$f(S) = \sum_{R \in S} \mathcal{C}(R) \quad (3)$$

while satisfying the following constraints:

$$\sum_{R \in S} x(e, R) = 1 \qquad \forall e \in E_R , \quad (4)$$

$$\mathcal{D}(R) \leq Q \qquad \forall R \in S . \quad (5)$$

The equality (4) means that each required edge must be serviced in precisely one of the routes, and the inequality (5) is the capacity constraint. Note that the CARP can also be formulated as an integer linear programming problem. Interested readers are referred to [4] or [8].

3   Deadheading Cycles

Even though a route can be represented by a sequence of serviced edges (with specified orientation), in reality a vehicle must travel in a continuous route, and so it may need to traverse some edges without servicing them when travelling between two required edges that are not physically adjacent. Such edges are called *deadheading edges*. In some cases, several deadheading edges may form a cycle, called a *deadheading cycle*. Indeed, provided it does not disconnect a route, a deadheading cycle can be removed without affecting feasibility of the route for two reasons: (i) the capacity constraint is still satisfied because serviced edges are unaffected (apart from potential reordering or reorientation), and (ii) with a route regarded as an Eulerian (multi)graph, removing a cycle preserves the parity of the degree of each vertex, so the Eulerian graph remains Eulerian after the removal.

It can be difficult to find a deadheading cycle while viewing a route as a sequence of serviced arcs because a deadheading cycle may be composed of deadheading edges which are traversed between different pairs of serviced arcs. So, for the purpose of detecting deadheading cycles, we view a route as an Eulerian graph (or multigraph if some edges are traversed more than once). Given a route $R = (a_1, a_2, \ldots, a_n)$, first we need to find shortest paths between v_0 and a_1, between a_i and a_{i+1} (for $i = 1, \ldots, n - 1$), and between a_n and v_0. This can be achieved by Dijkstra's algorithm [3]. Then, let G_{mult} be a multigraph such that the multiplicity of each edge in G_{mult} is equal to the number of times the edge is traversed (in any direction, with or without servicing) in R. When an edge is traversed three times or more, at least two such traversals are deadheading because, by definition, an edge can be serviced at most once. Every two deadheading traversals on the same edge correspond to a cycle in G_{mult}, which can be removed without disconnecting it as long as the number of traversals does not drop below 2. For an edge that is traversed just twice, careful consideration is needed before removing the corresponding cycle. In Fig. 1 for example, we can see that removing a deadheading cycle may or may not disconnect a route. Thus, to ensure continuity of a route, here we remove a deadheading cycle until the multiplicity of the corresponding edge reduces to either 1 or 2, depending on its parity.

After removing deadheading cycles, an updated sequence of edges traversed in the route R can be determined by finding an Eulerian cycle in the consequent G_{mult}. In order to obtain a representation of R as described in Sect. 2, if an edge to be serviced is traversed more than once, it is assumed that the edge is serviced at its first occurrence in the Eulerian cycle. Notice that the ordering and orientation of some serviced edges may change after the removal—see Fig. 1(a).

4   Description of the Tabu Search Algorithm

In this section, components of our tabu search algorithm are described. Our algorithm starts with an initial solution generated by the Path-Scanning algorithm [7]. This algorithm has been shown to produce better solutions on average than several other constructive algorithms [2].

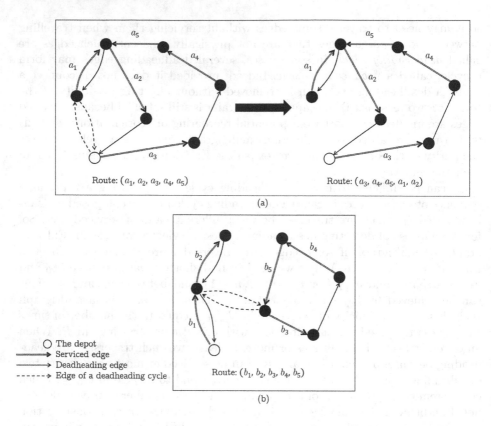

Fig. 1. Examples of deadheading cycles that are (a) removable and (b) not removable.

4.1 Neighbourhood Moves

Recall that a route is viewed as a sequence of required edges with specified directions that are serviced by the route. Four common neighbourhood moves are used:

1. *Single Insertion*: A required edge is removed from one route and then inserted at any position into another route. Both directions are examined when inserting it into a route.
2. *Double Insertion*: Two required edges are removed from one route and then inserted at any positions into another route. The edges may be inserted at the same position (i.e. between the same required edges, or between the depot and the first or last required edge). In the case where the edges are inserted at the same position, both possible permutations of the edges are considered ("edge 1 before edge 2" and "edge 2 before edge 1"). Both directions of each edge are examined when inserting it into a route. Thus, there are 8 possibilities of inserting two edges at a given position.

3. *Swap*: A required edge from each of two given routes is removed and inserted at any position into the other route, including the position of the removed edge. Both directions of each edge are examined when inserting it into a route.
4. *Two-Opt*: Each of two given routes is divided into two parts. Note that each part must have at least one required edge. Then, a part from one route is joined to a part from the other to create a new route. The remaining parts are also joined to create another route. Two possible ways of joining are examined. For clarity, this is illustrated in Fig. 2.

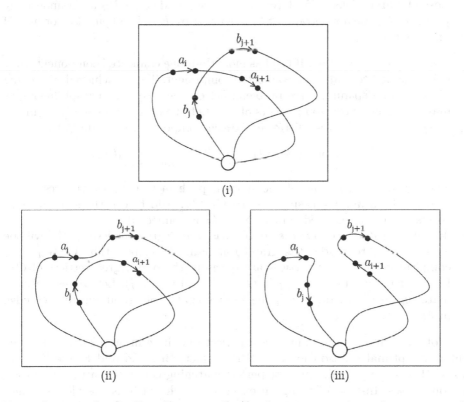

Fig. 2. An example of a Two-Opt move. In (i), after one route is cut between a_i and a_{i+1} and the other route between b_j and b_{j+1}, they can become either (ii) or (iii).

The first three neighbourhood moves were used in, for example, [2,9,12,15], and the Two-Opt move was utilised in, for example, [1,12,15].

4.2   Rural Postman Heuristic

When a Single Insertion, Double Insertion, or Swap move is implemented, an edge is inserted into a route without affecting the order of required edges already in the route. However, it is possible that reordering some of those edges might better accommodate the new edge and potentially lead to a better solution. Consider Single Insertion for example. Assume the move is feasible. Instead of specifying where to insert an edge into a route R, we may simply add it to the set of edges serviced by R and then attempt to construct a "promising" route that services this set of edges without having to keep the original order of any of the edges. In fact, we are attempting to solve a special (though still NP-hard) case of the CARP, namely the Rural Postman Problem (RPP), where the capacity is no less than the sum of required edges under consideration.

In our case, this reordering is achieved by means of a heuristic for the RPP proposed by Frederickson [5]. Given a set of required edges E_R in an underlying graph G, let G_R be a subgraph of G generated by E_R. The heuristic consists of two main steps:

1. *Connecting components*: If G_R has more than one connected component, they will be joined to make one connected component. This is achieved by solving the minimum spanning tree problem: Let G_S be a complete graph having as many vertices as the components of G_R. Let the cost of an edge $\{i, j\}$ in G_S be equal to the shortest distance between components C_i and C_j, that is,

$$\text{the cost of an edge } \{i, j\} \text{ in } G_S = \min_{u \in C_i, v \in C_j} d(u, v) \ ,$$

   where $d(u, v)$ is the cost of the shortest path in G between vertices u and v. Let T be a minimum spanning tree in G_S. Add to G_R the shortest path corresponding to each edge in T. Now G_R is connected.
2. *Matching odd-degree vertices*: If G_R has odd-degree vertices, paths will be added to G_R to render the graph Eulerian. To achieve this, let G_M be a complete graph whose vertices are precisely the odd-degree vertices of G_R. Let the cost of an edge $\{i, j\}$ in G_M be equal to $d(i, j)$. Let M be a perfect matching in G_M. Add to G_R the shortest path corresponding to each edge in M.

Note that finding a minimum-cost perfect matching in Step 2 does not guarantee an optimal solution for the RPP. In fact, Brandão and Eglese [2] have noted that using a minimum-cost perfect matching could give a worse solution in some cases. Instead of using an exact approach (such as the blossom algorithm [11]), we therefore opt to use a cheaper greedy method which operates by selecting the cheapest edge which is not adjacent to any previously selected edges.

Now G_R is an Eulerian (multi)graph. A new route corresponding to G_R can be obtained in the same way as we did for G_{mult} at the end of Sect. 3.

4.3   Tabu Record and Tabu Tenure

In our case, information on tabu moves is stored in arrays (as opposed to lists). The Single Insertion, Double Insertion and Swap moves share the same tabu array T_1. An entry $T_1(e, R)$ denotes the iteration number until which the insertion of the edge e into the route R is declared tabu (i.e. forbidden). Whenever a required edge e is removed from a route R, the entry $T_1(e, R)$ is updated so that $T_1(e, R)$ is set to the current iteration number plus the tabu tenure. Note that Double Insertion and Swap involve two insertions, so two entries in T_1 are updated. A Single Insertion/Double Insertion/Swap move is tabu if and only if all the insertions involved in the move are tabu. Note that the above procedure for updating T_1 still applies no matter whether the RPP heuristic is used because T_1 does not concern the position of an edge in a route.

Two-Opt uses a separate tabu array T_2. An entry $T_2(a, b)$ denotes the iteration number until which a cut (or, equivalently, a deadheading path) between required arcs a and b is declared tabu. (Recall that an arc is a directed edge.) Suppose a Two-Opt move involves a cut between required arcs a and b. Then, the entry $T_2(a, b)$ is updated so that $T_2(a, b)$ is set to the current iteration number plus the tabu tenure. It should be noted that a pair of arcs here must be treated as an ordered pair because different orders may correspond to different deadheading paths (see Fig. 3). In contrast, a pair (a, b) should be treated as identical to a pair $(-b, -a)$, where a minus sign denotes the opposite direction, because of symmetry of a route (as an undirected cycle in a graph). This can help save memory required for this tabu record.

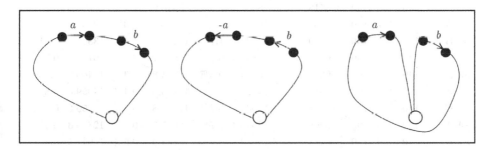

Fig. 3. Routes that contain the same required edges may be the same or different, depending on the direction and the order in which they are traversed.

For a tabu tenure, we follow a policy used in a previous tabu search algorithm in the literature [2]: the tenure is set to $n/2$, where n is the number of required edges. This remains fixed throughout the algorithm.

4.4   Admissibility of Moves

A common aspiration criterion is used here: In a given iteration, a neighbourhood move is considered if and only if it is feasible (the capacity constraint is satisfied) and either (1) it is non-tabu or (2) it is tabu but leads to a better solution than the current best solution.

Table 1. The mean and the coefficient of variation (CV) of solution costs for 20 independent runs on the EGL dataset

Instance	$|V|$	$|E|$	$|E_R|$	Best known solution	No reordering		RDC		RPP	
					Mean	CV	Mean	CV	Mean	CV
E1-A	77	51	98	3548	3548.0	0.0 %	3548.0	0.0 %	3548.0	0.0 %
E1-B	77	51	98	4498	4531.9	0.3 %	4525.4	0.2 %	4529.3	0.2 %
E1-C	77	51	98	5595	5748.3	1.2 %	5736.3	0.9 %	5634.1	0.9 %
E2-A	77	72	98	5018	5149.4	0.6 %	5059.8	0.4 %	5028.1	0.2 %
E2-B	77	72	98	6317	6347.5	0.1 %	6345.8	0.1 %	6339.3	0.1 %
E2-C	77	72	98	8335	8620.4	1.7 %	8683.6	1.8 %	8559.9	1.3 %
E3-A	77	87	98	5898	5916.1	0.3 %	5918.1	0.5 %	5937.4	0.7 %
E3-B	77	87	98	7775	8015.3	1.3 %	7912.9	1.2 %	7915.8	1.0 %
E3-C	77	87	98	10292	10392.1	0.6 %	10398.2	0.4 %	10371.4	0.3 %
E4-A	77	98	98	6444	6521.9	0.6 %	6492.8	0.4 %	6480.6	0.3 %
E4-B	77	98	98	8961	9135.3	1.1 %	9086.0	0.7 %	9029.7	0.3 %
E4-C	77	98	98	11550	11800.7	0.6 %	11795.6	0.6 %	11775.6	0.3 %
S1-A	140	75	190	5018	5132.7	0.9 %	5069.4	1.0 %	5077.5	1.1 %
S1-B	140	75	190	6388	6503.1	0.6 %	6480.1	1.0 %	6511.7	0.7 %
S1-C	140	75	190	8518	8662.8	0.9 %	8646.9	0.7 %	8623.6	0.8 %
S2-A	140	147	190	9884	10041.5	1.0 %	10089.1	1.0 %	10152.7	0.9 %
S2-B	140	147	190	13100	13531.4	1.1 %	13479.0	1.4 %	13366.8	0.9 %
S2-C	140	147	190	16425	16865.3	0.6 %	16888.1	0.7 %	16773.9	0.7 %
S3-A	140	159	190	10220	10414.8	0.8 %	10335.8	0.3 %	10460.9	0.9 %
S3-B	140	159	190	13682	14126.6	1.5 %	13935.6	0.4 %	13959.0	0.8 %
S3-C	140	159	190	17188	17634.8	0.7 %	17490.8	0.5 %	17456.6	0.4 %
S4-A	140	190	190	12268	12527.3	0.7 %	12555.5	0.7 %	12775.3	0.5 %
S4-B	140	190	190	16283	16613.7	0.7 %	16534.9	0.5 %	16641.5	0.5 %
S4-C	140	190	190	20481	21012.5	0.5 %	20980.8	0.7 %	21059.6	0.4 %
Average				9736.9	9949.7		9916.2**		9917.0	

5   Computational Results

The tabu search algorithm described above was coded using C++ and all experiments were performed on Intel Core i3-2120 3.30 GHz with 8 GB RAM using benchmark datasets EGL, BMCV, and EGL-Large.[1] To simulate a real-life scenario where speed is preferable to optimality, a time limit of 300 s was introduced for the EGL and BMCV sets. Due to a larger number of required edges, a longer time limit of 1,200 s was introduced for the EGL-Large set.

To investigate how the edge-reordering procedures might help to improve the search, three versions of the algorithms were tested: the first one does not

---

[1] These datasets, as well as best known solutions, are available at http://logistik.bwl. uni-mainz.de/benchmarks.php.

Table 2. The mean and the coefficient of variation (CV) of solution costs for 20 independent runs on the BMCV dataset

Subset of instances	Average over 25 instances in the subset						
	Best known solution	No Reordering		RDC		RPP	
		Mean	CV	Mean	CV	Mean	CV
C (C1 - C25)	3683.4	3781.1	1.5%	3760.9	1.2%	3747.1	1.1%
D (D1 - D25)	2872.4	2936.4	1.0%	2896.8	0.9%	2892.7	0.8%
E (E1 - E25)	3698.0	3795.9	1.5%	3782.0	1.6%	3774.8	1.2%
F (F1 - F25)	3003.0	3058.4	0.9%	3035.8	1.0%	3026.7	0.7%
Average	3314.2	3392.9		3368.9**		3360.3**	

Table 3. The mean and the coefficient of variation (CV) of solution costs for 20 independent runs on the EGL-Large dataset

Instance	—V—	—E—	$|E_R|$	Best known solution	No reordering		RDC		RPP	
					Mean	CV	Mean	CV	Mean	CV
G1-A	255	347	375	1004864	1025967.0	0.6%	1024128.5	0.6%	1046700.8	0.8%
G1-B	255	347	375	1129937	1135307.6	0.4%	1134844.3	0.4%	1154564.5	0.7%
G1-C	255	347	375	1262888	1282483.9	0.6%	1279901.8	0.7%	1303527.5	0.5%
G1-D	255	347	375	1398958	1410589.4	0.7%	1408335.1	0.6%	1423174.4	0.5%
G1-E	255	347	375	1543804	1551973.2	0.9%	1550111.0	0.8%	1581999.2	0.7%
G2-A	255	375	375	1115339	1120506.3	0.5%	1120201.9	0.6%	1142626.4	0.6%
G2-B	255	375	375	1226645	1237201.4	0.7%	1235783.5	0.7%	1281057.4	0.5%
G2-C	255	375	375	1371004	1381432.9	0.6%	1378903.3	0.7%	1426369.6	0.8%
G2-D	255	375	375	1509990	1515175.8	0.4%	1511281.5	0.4%	1557018.8	0.8%
G2-E	255	375	375	1659217	1667802.9	0.6%	1663952.9	0.6%	1696421.8	0.5%
Average				1322264.6	1332844.0		1330744.3**		1361346.0**	

consider deadheading cycles nor the RPP heuristic ("No Reordering"), and each of the other two implements one of the procedures:

1. attempting to remove deadheading cycles from an initial solution and from routes that are affected by a neighbourhood move in each iteration ("RDC"),
2. integrating the RPP heuristic within the neighbourhood moves ("RPP").

For each version, 20 independent runs were conducted on each instance.

Tables 1, 2 and 3 present the features of the instances together with the mean and the coefficient of variation of the best solution costs from 20 independent runs of each version of our algorithm as described above. Due to a large number (100) of instances in the BMCV dataset, each row of Table 2 shows the average results on a subset of 25 instances

Tables 1 and 2 suggest that both procedures improve the quality of solutions produced compared to the "No Reordering" version (here, the ** symbol indicates statistical significance according to a Wilcoxon Signed Rank test with

$p < 0.01$). However, for the EGL-Large dataset in particular, their performances are noticeably different. Table 3 shows that attempting to remove deadheading cycles can improve a solution on all EGL-Large instances, while the RPP heuristic does not seem to give any improvement. This is very likely because the RPP heuristic requires a long computational time for large instances, as can be seen in Table 4.

Table 4. The mean and the coefficient of variation (CV) of the time taken (seconds) by the edge-reordering procedures for 20 independent runs on the EGL-Large dataset

Instance	RDC		RPP	
	Mean	CV	Mean	CV
G1-A	12.8	2.6 %	1171.9	0.2 %
G1-B	17.2	2.5 %	1165.7	0.1 %
G1-C	19.4	2.4 %	1161.0	0.1 %
G1-D	20.7	3.2 %	1152.6	0.2 %
G1-E	22.0	1.5 %	1139.2	0.2 %
G2-A	14.4	1.9 %	1164.9	0.1 %
G2-B	15.3	4.2 %	1165.9	0.1 %
G2-C	17.6	4.3 %	1158.2	0.2 %
G2-D	18.7	3.6 %	1149.4	0.5 %
G2-E	19.0	1.9 %	1143.1	0.5 %
Average	17.7		1157.2	

Table 4 shows the mean and coefficient of variation of the time taken by each edge-reordering procedure in the respective version of the algorithm. As no optimal solutions are known for this dataset, the algorithm halts once the time limit of 1,200 s has elapsed. Table 4 shows that the RPP heuristic uses the vast majority of computation time (about 96 %). This could be because the RPP heuristic is integrated with all Single Insertion, Double Insertion and Swap moves, comprising a huge set of neighbour solutions to consider with the heuristic. Using a large amount of time to explore a neighbourhood means this version of the algorithm performed considerably fewer iterations. As a result, the search may have not moved "far" from the initial solution in the solution space.

It is clear from Tables 1, 2 and 3 that solutions from all versions of the algorithm in this work are still far from the best known solutions. Nevertheless, we have seen that re-ordering edges has potential to guide the search to a promising region of the solution space and obtain better solutions within the same amount of time.

6   Conclusion and Future Work

This paper brought to attention two ideas to help guide a tabu search or, in fact, any local search method to a promising region of the solution space for the Capacitated Arc Routing Problem. The first idea is to investigate deadheading cycles and attempt to remove them after generating an initial solution and when they appear as a result of neighbourhood moves. Removing deadheading cycles guarantees an improvement, provided all edge costs are non-zero. Nevertheless, it is important to note that some deadheading cycles might not be removable as doing so may disconnect a route.

One way to ensure the continuity of a route is to remove a deadheading cycle only if the multiplicity of the corresponding edge does not drop below 2. However, we might try removing each possible deadheading cycle and directly checking if the route still remains connected. This allows us to detect more removable deadheading cycles, but it is important to find an efficient algorithm to do so.

Moreover, this work considered only deadheading cycles that result from a single edge traversed repeatedly. There can also be a cycle composed of several deadheading edges that are traversed precisely once. Still, an efficient algorithm is required for detecting the removability of such cycle.

This paper also investigated a combination of a heuristic for the Rural Post-man problem with neighbourhood moves. This allows edges in a route to be re-ordered and potentially gives a better solution that might normally require several traditional neighbourhood moves. This can increase connectivity of the solution space and, given excess time, increase the probability of reaching good solutions.

Experimental results showed that both ideas have potential to improve a search. However, it can be time-consuming to try the Rural Postman heuristic with all Single Insertion, Double Insertion, and Swap moves. It would therefore be interesting to find a balance between increasing connectivity of the solution space and taking time to evaluate all neighbours. Moreover, these two ideas may not be effective alone as can be seen from a comparison between solutions from the algorithm in this work and best known solutions. One may try to combine these two ideas together rather than use them separately, or even combine them with traditional tabu search techniques such as intensification and diversification. Such a good combination is still to be researched.

References

1. Beullens, P., Muyldermans, L., Cattrysse, D., Van Oudheusden, D.: A guided local search heuristic for the capacitated arc routing problem. Eur. J. Oper. Res. **147**(3), 629–643 (2003)
2. Brandão, J., Eglese, R.: A deterministic tabu search algorithm for the capacitated arc routing problem. Comput. Oper. Res. **35**(4), 1112–1126 (2008)
3. Dijkstra, E.W.: A note on two problems in connexion with graphs. Numerische Mathematik **1**(1), 269–271 (1959)

4. Eiselt, H.A., Gendreau, M., Laporte, G.: Arc routing problems, Part II: the rural postman problem. Oper. Res. **43**(3), 399–414 (1995)
5. Frederickson, G.N.: Approximation algorithms for some postman problems. J. ACM (JACM) **26**(3), 538–554 (1979)
6. Fu, H., Mei, Y., Tang, K., Zhu, Y.: Memetic algorithm with heuristic candidate list strategy for capacitated arc routing problem. In: 2010 IEEE Congress on Evolutionary Computation (CEC), pp. 1–8. IEEE (2010)
7. Golden, B.L., DeArmon, J.S., Baker, E.K.: Computational experiments with algorithms for a class of routing problems. Comput. Oper. Res. **10**(1), 47–59 (1983)
8. Golden, B.L., Wong, R.T.: Capacitated arc routing problems. Networks **11**(3), 305–315 (1981)
9. Greistorfer, P.: A tabu scatter search metaheuristic for the arc routing problem. Comput. Ind. Eng. **44**(2), 249–266 (2003)
10. Hertz, A., Laporte, G., Mittaz, M.: A tabu search heuristic for the capacitated arc routing problem. Oper. Res. **48**(1), 129–135 (2000)
11. Kolmogorov, V.: Blossom V: a new implementation of a minimum cost perfect matching algorithm. Math. Program. Comput. **1**(1), 43–67 (2009)
12. Lacomme, P., Prins, C., Ramdane-Cherif, W.: Competitive memetic algorithms for arc routing problems. Ann. Oper. Res. **131**(1–4), 159–185 (2004)
13. Polacek, M., Doerner, K.F., Hartl, R.F., Maniezzo, V.: A variable neighborhood search for the capacitated arc routing problem with intermediate facilities. J. Heuristics **14**(5), 405–423 (2008)
14. Santos, L., Coutinho-Rodrigues, J., Current, J.R.: An improved ant colony optimization based algorithm for the capacitated arc routing problem. Transp. Res. Part B Methodol. **44**(2), 246–266 (2010)
15. Tang, K., Mei, Y., Yao, X.: Memetic algorithm with extended neighborhood search for capacitated arc routing problems. IEEE Trans. Evol. Comput. **13**(5), 1151–1166 (2009)

Multi-chaotic Approach for Particle Acceleration in PSO

Michal Pluhacek[1(✉)], Roman Senkerik[1], Adam Viktorin[1],
and Ivan Zelinka[2]

[1] Faculty of Applied Informatics, Tomas Bata University in Zlin,
Nam T.G. Masaryka 5555, 760 01 Zlin, Czech Republic
{pluhacek,senkerik,aviktorin}@fai.utb.cz
[2] Faculty of Electrical Engineering and Computer Science, Technical University
of Ostrava, 17. Listopadu 15, 708 33 Ostrava-Poruba, Czech Republic
ivan.zelinka@vsb.cz

Abstract. This paper deals with novel approach for hybridization of two scientific techniques: the evolutionary computational techniques and deterministic chaos. The Particle Swarm Optimization algorithm is enhanced with two pseudo-random number generators based on chaotic systems. The chaotic pseudo-random number generators (CPRNGs) are used to guide the particles movement through multiplying the accelerating constants. Different CPRNGs are used simultaneously in order to improve the performance of the algorithm. The IEEE CEC'13 benchmark suite is used to test the performance of the proposed method.

Keywords: Particle swarm optimization · PSO · Chaos · Acceleration constant

1 Introduction

The Particle Swarm Optimization algorithm (PSO) [1–4] is a highly popular meta-heuristic for complex optimization tasks. It has been shown in [5, 6] that various evolutionary computational techniques (ECTs) can benefit from implementation of chaotic sequences. The chaotic PSO was proposed in [7] and it has been shown that the performance of the PSO can be significantly improved when the chaotic sequences are implemented as pseudo-random number generators (PRNGs) for the PSO. Successful applications of chaos PSO were presented among others in [8, 9].

More recently the utilization of more than one chaotic PRNG (CPRNG) during single run of PSO was proposed in [10] showing promising results. This work builds on the idea presented in [10] and presents a different approach for the hybridization of PSO and multiple chaotic systems.

The novelty of the proposed approach is that the pair of random numbers needed for the main PSO formula is generated using a pair of different chaotic pseudo-random number generators.

The following research questions were defined:

(1) Will the performance of PSO algorithm change when the pair of CPRNGs is implemented into the main PSO formula?

© Springer International Publishing Switzerland 2016
M.J. Blesa et al. (Eds.): HM 2016, LNCS 9668, pp. 75–86, 2016.
DOI: 10.1007/978-3-319-39636-1_6

(2) Is it possible to improve the performance of PSO algorithm by this approach?

(3) Is the performance of PSO with pair of CPRNGs better than the performance of PSO with single CPRNG?

The paper is structured as follows: brief description of the PSO algorithm is presented in Sect. 2, the description of utilized chaotic systems (maps) is given in Sect. 3. The experiment setup is detailed in Sect. 4 and results presented in Sect. 5. The paper concludes with a discussion of the results and future research.

2   Particle Swarm Optimization Algorithm with Two CPRNGs

Originally the PSO was inspired by behavior of fish schools and bird flocks. The knowledge of global best found solution (typically noted *gBest*) is shared among the particles in the swarm. Furthermore each particle has the knowledge of its own (personal) best found solution (noted *pBest*). Last important part of the algorithm is the velocity of each particle that is taken into account during the calculation of the particle movement. The new position of each particle is then given by (1), where x_i^{t+1} is the new particle position; x_i^t refers to current particle position and v_i^{t+1} is the new velocity of the particle.

$$x_i^{t+1} = x_i^t + v_i^{t+1} \tag{1}$$

To calculate the new velocity the distance from *pBest* and *gBest* is taken into account alongside with current velocity (2).

$$v_{ij}^{t+1} = w \cdot v_{ij}^t + c_1 \cdot Rand_1 \cdot (pBest_{ij} - x_{ij}^t) + c_2 \cdot Rand_2 \cdot (gBest_j - x_{ij}^t) \tag{2}$$

Where:

v_{ij}^{t+1} - New velocity of the ith particle in iteration $t + 1$. (component j of the dimension D).

w – Inertia weight value.

v_{ij}^t - Current velocity of the ith particle in iteration t. (component j of the dimension D).

c_1, c_2 - Acceleration constants.

$pBest_{ij}$ – Local (personal) best solution found by the ith particle. (component j of the dimension D).

$gBest_j$ - Best solution found in a population. (component j of the dimension D).

x_{ij}^t - Current position of the ith particle (component j of the dimension D) in iteration t.

$Rand_1$ – Pseudo random number, interval (0, 1). Generated by first chaotic PRNG.

$Rand_2$ – Pseudo random number, interval (0, 1). Generated by second chaotic PRNG.

In this work two different chaotic PRNGs are used to generate the pseudo-random numbers in the velocity calculation equation. In this way each generator has direct effect on the trajectory of the particle regarding its tendencies to move towards the *pBest* or *gBest*.

Finally the linear decreasing inertia weight [2, 3] is used. The dynamic inertia weight is meant to slow the particles over time in order to improve the local search capability in the later phase of the optimization. The inertia weight has two control parameters w_{start} and w_{end}. A new w for each iteration is given by (3), where t stands for current iteration number and n stands for the total number of iterations. The typical values used in this study were $w_{start} = 0.9$ and $w_{end} = 0.4$.

$$w = w_{start} - \frac{((w_{start} - w_{end}) \cdot t)}{n} \tag{3}$$

3 Chaotic Maps

In this section six discrete dissipative chaotic systems (maps) are described. These six chaotic maps were used as CPRNG's for the process of new velocity calculation in PSO (See (2)). The selection of this particular set of maps was based on previous experiments.

3.1 Lozi Chaotic Map

The Lozi map is a simple discrete two-dimensional chaotic map. The map equations are given in (4). The typical parameter values are: a = 1.7 and b = 0.5 with respect to [20]. For these values, the system exhibits typical chaotic behavior and with this parameter setting it is used in the most research papers and other literature sources. The x, y plot of Lozi map with the typical setting is depicted in Fig. 1.

$$\begin{aligned} X_{n+1} &= 1 - a|X_n| + bY_n \\ Y_{n+1} &= X_n \end{aligned} \tag{4}$$

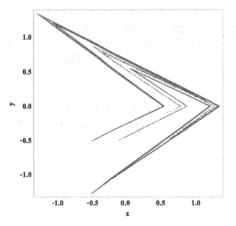

Fig. 1. x,y plot of Lozi map

3.2 Dissipative Standard Map

The Dissipative standard map is a two-dimensional chaotic map [11]. The parameters used in this work are $b = 0.6$ and $k = 8.8$ based on previous experiments [15, 16] and suggestions in literature [20]. The x, y plot of Dissipative standard map is given in Fig. 2. The map equations are given in (5).

$$X_{n+1} = X_n + Y_{n+1}(\mathrm{mod}2\pi)$$
$$Y_{n+1} = bY_n + k\sin X_n(\mathrm{mod}2\pi)$$

(5)

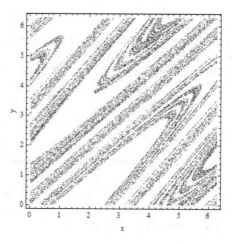

Fig. 2. x,y plot of Dissipative standard map

3.3 Arnold's Cat Map

The Arnold's Cat map is a simple two dimensional discrete system that stretches and folds points (x, y) to $(x + y, x + 2y)$ mod 1 in phase space. The map equations are given in (6). This map was used with parameter $k = 0.1$. The visualization of the map is given in Fig. 3.

$$X_{n+1} = X_n + Y_n(\mathrm{mod}1)$$
$$Y_{n+1} = X_n + kY_n(\mathrm{mod}1)$$

(6)

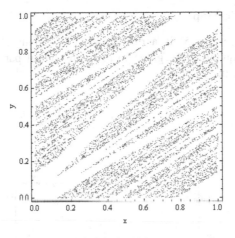

Fig. 3. Arnold' Cat chaotic map – x,y plot

3.4 Sinai Map

The Sinai map (Fig. 4) is a simple two dimensional discrete system similar to the Arnold's Cat map. The map equations are given in (7). The parameter used in this work is $\delta = 0.1$ as suggested in [18].

$$X_{n+1} = X_n + Y_n + \delta \cos 2\pi Y_n (\mathrm{mod}1)$$
$$Y_{n+1} = X_n + 2Y_n (\mathrm{mod}1)$$

$$(7)$$

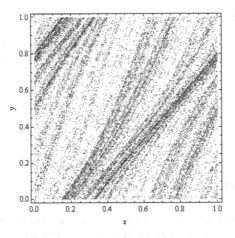

Fig. 4. Sinai chaotic map – x, y plot

3.5   Burgers Chaotic Map

The Burgers map (See Fig. 5) is a discretization of a pair of coupled differential equations The map equations are given in (8) with control parameters $a = 0.75$ and $b = 1.75$ as suggested in [11].

$$X_{n+1} = aX_n - Y_n^2$$
$$Y_{n+1} = bY_n + X_nY_n$$

(8)

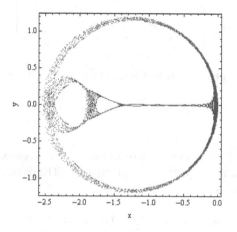

Fig. 5. x,y plot of Burgers map

3.6   Tinkerbell Map

The Tinkerbell map is a two-dimensional complex discrete-time dynamical system given by (9) with following control parameters: $a = 0.9$, $b = -0.6$, $c = 2$ and $d = 0.5$ [11]. The x,y plot of the Tinkerbell map is given in Fig. 6.

$$X_{n+1} = X_n^2 - Y_n^2 + aX_n + bY_n$$
$$Y_{n+1} = 2X_nY_n + cX_n + dY_n$$

(9)

4   Experiment Setup and Results

In this work the performance of the newly proposed hybrid method was tested on the IEEE CEC 2013 benchmark set [12] for dimension setting (*dim*) = 10. The benchmark suite contains a total of 28 functions (noted hereinafter $f_1 - f_{28}$).

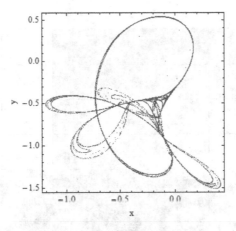

Fig. 6. x,y plot of Tinkerbell map

According to the benchmark rules 51 separate runs were performed for each algorithm and the maximum number of cost function evaluations (CFE) was set to 100000. The population size was set to 40. Other controlling parameters of the PSO were set to typical [1–4] values as follows:

c_1, $c_2 = 2$; $w_{start} = 0.9$; $w_{end} = 0.4$; $v_{max} = 0.2$; For these values the PSO exhibits typical behavior and the proposed method is not oversensitive to the setting of these parameters.

In the first experiment all possible combinations of CPRNGs based on six above described chaotic systems were tested on selected functions from the benchmark suite.

The notation corresponds to above presented subsections structure. The Lozi map based PRNG is noted hereinafter as "1". Dissipative standard map as "2", Arnold´s Cat map as "3", Sinai Map as "4", Burgrs chaotic map as "5" and Tinkerbell map as "6".

Mean results comparison from 51 repeated runs of the algorithm for each setting are given in heat maps presented in Figs. 7, 8, 9, 10 and 11. The best mean result achieved is highlighted by black cell background, the worst mean results is given by white cell background. "The column number corresponds to the CPRNG generating $Rand_1$, while the row number shows the CPRNG that generates $Rand_2$. For example: the best result for f_4 was achieved by combination of Sinai map based CPRNG and Tinkerbell map based CPRNG noted hereinafter as "6-3".

The data suggest that it may be beneficial to use a pair of different chaotic systems rather than a single chaotic system (as the main diagonal in the tables contains worse results than the best results).

Based on the results presented above four promising combinations of CPRNGs were chosen and the performance of the algorithms were tested using whole CEC13 Benchmark suite. The results of this second experiment are presented in Table 1. Best

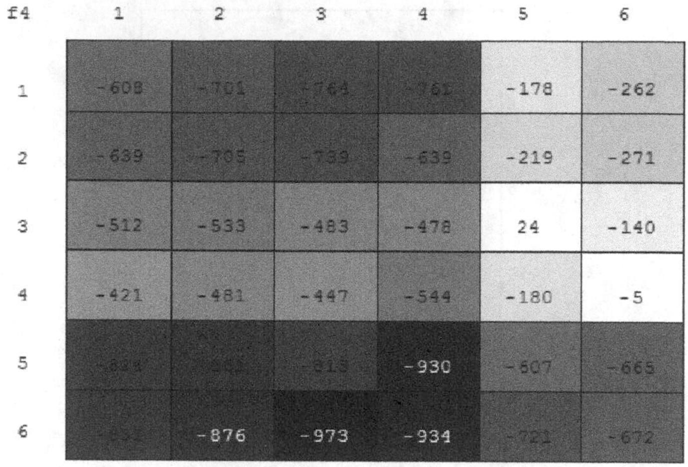

f4	1	2	3	4	5	6
1	-608	-701	-764	-761	-178	-262
2	-639	-705	-739	-639	-219	-271
3	-512	-533	-483	-478	24	-140
4	-421	-481	-447	-544	-180	-5
5	-888	-865	-813	-930	-807	-665
6	-932	-876	-973	-934	-721	-672

Fig. 7. Mean results (rounded) comparison f_4

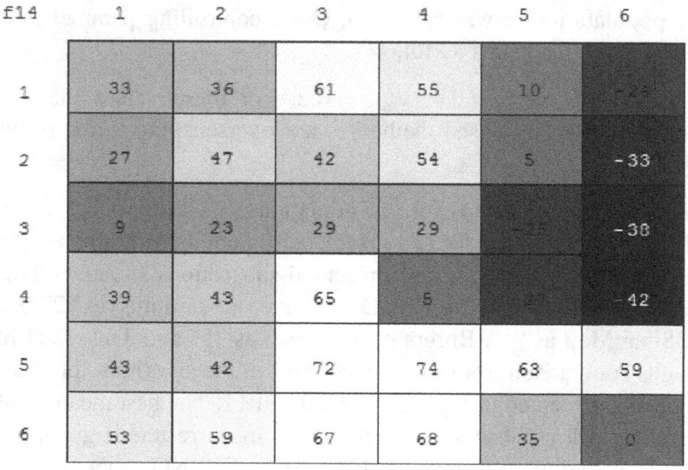

f14	1	2	3	4	5	6
1	33	36	61	55	10	-24
2	27	47	42	54	5	-33
3	9	23	29	29	-25	-38
4	39	43	65	6	-4	-42
5	43	42	72	74	63	59
6	53	59	67	68	35	0

Fig. 8. Mean results (rounded) comparison f_{14}

results are given in bold numbers (if significant difference can be identified using standard statistics). The results of multi-chaotic versions of PSO algorithm are compared with the canonical (non-chaotic) PSO algorithm (noted as PSO).

In the Table 1 the benchmark functions are divided into unimodal (noted with ^{u}), basic multimodal (noted with ^{m}) and composite functions (noted with ^{c}). According to the data presented in Table 1 it seems that it is possible to significantly improve the

f15
1 2 3 4 5 6
1 793 809 764 847 811
2 787 815 806 863 837
3 812 800 836 936 816
4 792 798 783 934 857
5 824 820 711 883
6 793 778 728 822 783

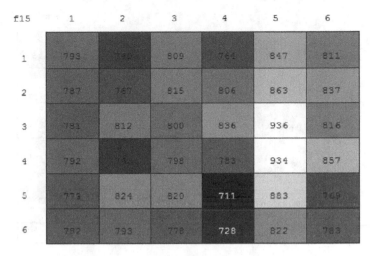

Fig. 9. Mean results (rounded) comparison f_{15}

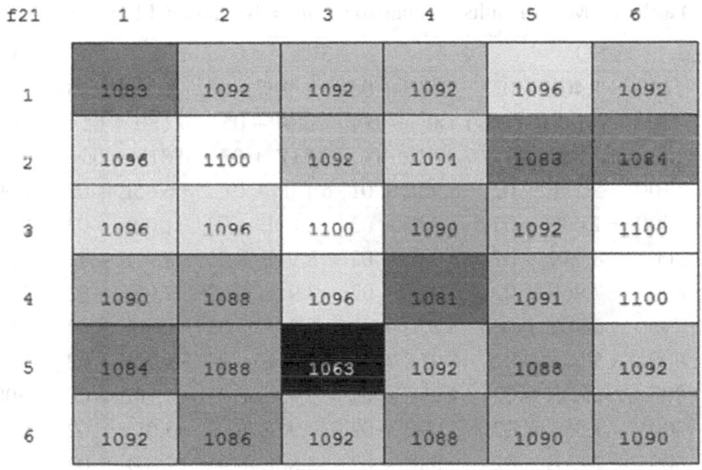

f21
1 2 3 4 5 6
1 1083 1092 1092 1092 1096 1092
2 1096 1100 1092 1091 1083 1084
3 1096 1096 1100 1090 1092 1100
4 1090 1088 1096 1081 1091 1100
5 1084 1088 1063 1092 1088 1092
6 1092 1086 1092 1088 1090 1090

Fig. 10. Mean results (rounded) comparison f_{21}

performance of PSO algorithm on certain problems by implementation of a pair of CPRNGs. In the case of unimodal functions ($f_1 - f_5$) it is possible to obtain significantly better results or optima when proper combination of CPRNG is used. Further the performance if significantly better in the case of f_{14}, f_{15} and f_{22}.

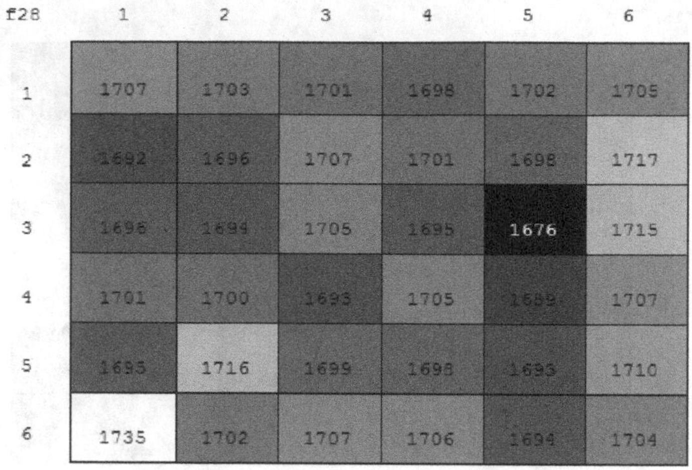

Fig. 11. Mean results (rounded) comparison f_{28}

Table 1. Mean results comparison, dim = 10, max. CFE = 100000

Function	f_{min}	PSO	4–6	3–5	1–2	6–3
f_1^u	−1400	−1.40E + 03	−1.40E + 03	−1.40E + 03	−1.40E + 03	−1.40E + 03
f_2^u	−1300	2.45E + 05	**1.18E + 05**	1.28E + 05	4.18E + 05	2.76E + 05
f_3^u	−1200	1.86E + 06	7.49E + 05	**2.71E + 05**	8.81E + 05	1.37E + 06
f_4^u	−1100	−5.20E + 02	−5.92E + 01	6.15E + 01	−6.65E + 02	**−9.39E + 02**
f_5^u	−1000	−1.00E + 03	−1.00E + 03	−1.00E + 03	−1.00E + 03	−1.00E + 03
f_6^m	−900	−8.94E + 02	−8.93E + 02	−8.94E + 02	−8.91E + 02	−8.91E + 02
f_7^m	−800	−7.96E + 02	−7.97E + 02	−7.98E + 02	−7.95E + 02	−7.92E + 02
f_8^m	−700	−6.80E + 02	−6.80E + 02	−6.80E + 02	−6.80E + 02	−6.80E + 02
f_9^m	−600	−5.97E + 02	−5.97E + 02	−5.97E + 02	−5.96E + 02	−5.96E + 02
f_{10}^m	−500	−5.00E + 02	−4.99E + 02	−4.99E + 02	−4.99E + 02	−5.00E + 02
f_{11}^m	−400	−3.98E + 02	−3.99E + 02	−3.98E + 02	−3.98E + 02	−3.97E + 02
f_{12}^m	−300	−2.87E + 02	−2.88E + 02	−2.88E + 02	−2.87E + 02	−2.85E + 02
f_{13}^m	−200	−1.80E + 02	−1.85E + 02	−1.83E + 02	−1.82E + 02	−1.74E + 02
f_{14}^m	−100	5.72E + 01	**−1.90E + 01**	9.42E − 01	3.03E + 01	7.50E + 01
f_{15}^m	100	8.45E + 02	9.59E + 02	9.44E + 02	**7.58E + 02**	7.96E + 02
f_{16}^m	200	2.01E + 02	2.01E + 02	2.01E + 02	2.01E + 02	2.01E + 02
f_{17}^m	300	3.14E + 02	3.13E + 02	3.14E + 02	3.13E + 02	3.14E + 02
f_{18}^m	400	4.32E + 02	4.30E + 02	4.33E + 02	4.23E + 02	4.24E + 02
f_{19}^m	500	5.01E + 02	5.01E + 02	5.01E + 02	5.01E + 02	5.01E + 02
f_{20}^m	600	6.03E + 02	6.02E + 02	6.03E + 02	6.02E + 02	6.03E + 02
f_{21}^c	700	1.08E + 03	1.09E + 03	1.09E + 03	1.09E + 03	1.08E + 03
f_{22}^c	800	9.72E + 02	**8.98E + 02**	9.43E + 02	9.87E + 02	1.04E + 03

(Continued)

Table 1. (*Continued*)

Function	f_{min}	PSO	4−6	3−5	1−2	6−3
f_{23}^c	900	1.81E + 03	1.78E + 03	1.73E + 03	1.77E + 03	1.86E + 03
f_{24}^c	1000	1.21E + 03	1.20E + 03	1.20E + 03	1.21E + 03	1.21E + 03
f_{25}^c	1100	1.30E + 03	1.30E + 03	1.30E + 03	1.31E + 03	1.31E + 03
f_{26}^c	1200	1.36E + 03	1.34E + 03	1.33E + 03	1.37E + 03	1.35E + 03
f_{27}^c	1300	1.67E + 03	1.66E + 03	1.67E + 03	1.67E + 03	1.68E + 03
f_{28}^c	1400	1.69E + 03	1.71E + 03	1.67E + 03	1.71E + 03	1.70E + 03

5 Conclusion

In this work a pair of different chaotic pseudo-random number generators was utilized for generating the pair of random numbers for the main formula of PSO algorithm. In the experimental part six different chaotic systems were utilized as CPRNGs. The performance of all possible pairs of CPRNG was tested on 5 selected functions from the IEEE CEC'13 Benchmark set. The results suggest that it may be beneficial to use a pair of different chaotic systems rather than a single chaotic system.

Several promising pairs of CPRNGs were implemented into PSO and the performance of proposed algorithms was tested on IEEE CEC'13 benchmark suite and compared with the performance of canonical version of the algorithm.

Regarding the research questions following statements apply:

(1) Yes, the performance of the algorithm changes when different pairs of chaotic pseudo-random generators are applied.
(2) Yes, it is very likely that the performance of PSO algorithm on a particular problem can be improved when a proper pair of chaotic PRNGs is implemented in the way described in this paper.
(3) Yes, as has been pointed out above it seems very likely that the performance of PSO algorithm with a pair of CPRNGs may be better than the performance of PSO with a single CPRNG (on the premise that a tuning experiment is performed in order to select the best combination of CPRNGs)

In this research a novel approach for multi-chaotic PSO was proposed and tested. The results suggest that the chaotic sequences can improve the performance of evolutionary computational techniques. The research of the mutual interaction of chaos and ECTs remains the main focus for our future research. Based on the study presented in this work an adaptive approach for selection of the best pair of CPRNs will be designed in the future.

Acknowledgements. This work was supported by Grant Agency of the Czech Republic – GACR P103/15/06700S, further by the Ministry of Education, Youth and Sports of the Czech Republic within the National Sustainability Programme Project no. LO1303 (MSMT-7778/2014). Also by the European Regional Development Fund under the Project CEBIA-Tech no. CZ.1.05/ 2.1.00/03.0089 and by Internal Grant Agency of Tomas Bata University under the Project no. IGA/Ceb-iaTech/2016/007.

References

1. Kennedy, J., Eberhart, R.: Particle swarm optimization. In: Proceedings of IEEE International Conference on Neural Networks, vol. IV, pp. 1942–1948 (1995)
2. Shi, Y.H., Eberhart, R.C.: A modified particle swarm optimizer. In: IEEE International Conference on Evolutionary Computation, pp. 69–73, Anchorage Alaska (1998)
3. Nickabadi, M., Ebadzadeh, M., Safabakhsh, R.: A novel particle swarm optimization algorithm with adaptive inertia weight. Appl. Soft Comput. **11**(4), 3658–3670 (2011). ISSN 1568-4946
4. Eberhart, R., Kennedy, J.: Swarm Intelligence. The Morgan Kaufmann Series in Artificial Intelligence. Morgan Kaufmann, Los Altos (2001)
5. Caponetto, R., Fortuna, L., Fazzino, S., Xibilia, M.G.: Chaotic sequences to improve the performance of evolutionary algorithms. IEEE Trans. Evol. Comput. **7**(3), 289–304 (2003)
6. Davendra, D., Zelinka, I., Senkerik, R.: Chaos driven evolutionary algorithms for the task of PID control. Comput. Math Appl. **60**(4), 1088–1104 (2010). ISSn 0898-1221
7. Alatas, B., Akin, E., Ozer, B.A.: Chaos embedded particle swarm optimization algorithms. Chaos, Solitons Fractals **40**(4), 1715–1734 (2009). ISSN 0960-0779
8. Pluhacek, M., Senkerik, R., Davendra, D., Zelinka, I.: Designing PID controller for DC motor system by means of enhanced PSO algorithm with discrete chaotic Lozi map. In: Proceedings of the 26th European Conference on Modelling and Simulation, ECMS 2012, pp. 405–409 (2012). ISBN 978-0-9564944-4-3
9. Araujo, E., Coelho, L.: Particle swarm approaches using Lozi map chaotic sequences to fuzzy modelling of an experimental thermal-vacuum system. Appl. Soft Comput. **8**(4), 1354–1364 (2008)
10. Pluhacek, M., Senkerik, R., Davendra, D., Zelinka, I.: Particle swarm optimization algorithm driven by multichaotic number generator. Soft. Comput. **18**(4), 631–639 (2014)
11. Sprott, J.C.: Chaos and Time-Series Analysis. Oxford University Press, Oxford (2003)
12. Liang, J.J., Qu, B.-Y., Suganthan, P.N., Hernández-Díaz, A.G.: Problem definitions and evaluation criteria for the CEC 2013 special session and competition on real-parameter optimization. Technical report 201212, Computational Intelligence Laboratory, Zhengzhou University, Zhengzhou China and Technical report, Nanyang Technological University, Singapore (2013)

Districting and Routing for Security Control

Michael Prischink[1,2], Christian Kloimüllner[1(✉)], Benjamin Biesinger[1],
and Günther R. Raidl[1]

[1] Institute of Computer Graphics and Algorithms, TU Wien,
Favoritenstraße 9–11/1861, 1040 Vienna, Austria
{kloimuellner,biesinger,raidl}@ac.tuwien.ac.at
[2] Research Industrial Systems Engineering, Concorde Business Park F,
2320 Schwechat, Austria
michael.prischink@rise-world.com

Abstract. Regular security controls on a day by day basis are an essential and important mechanism to prevent theft and vandalism in business buildings. Typically, security workers patrol through a set of objects where each object requires a particular number of visits on all or some days within a given planning horizon, and each of these visits has to be performed in a specific time window. An important goal of the security company is to partition all objects into a minimum number of disjoint clusters such that for each cluster and each day of the planning horizon a feasible route for performing all the requested visits exists. Each route is limited by a maximum working time, must satisfy the visits' time window constraints, and any two visits of one object must be separated by a minimum time difference. We call this problem the *Districting and Routing Problem for Security Control*. In our heuristic approach we split the problem into a districting part where objects have to be assigned to districts and a routing part where feasible routes for each combination of district and period have to be found. These parts cannot be solved independently though. We propose an exact mixed integer linear programming model and a *routing construction heuristic* in a greedy like fashion with *variable neighborhood descent* for the routing part as well as a *districting construction heuristic* and an *iterative destroy & recreate* algorithm for the districting part. Computational results show that the exact algorithm is only able to solve small routing instances and the iterative destroy & recreate algorithm is able to reduce the number of districts significantly from the starting solutions.

1 Introduction

As in the area of private security control constant surveillance of an object might not be economically viable or even necessary, security firms face the problem of

We want to thank Günter Kiechle and Fritz Payr from CAPLAS GmbH for the collaboration on this topic. This work is supported by the Austrian Science Fund (FWF) grant P24660-N23 and by the Austrian Research Promotion Agency (FFG) under contract 849028.

© Springer International Publishing Switzerland 2016
M.J. Blesa et al. (Eds.): HM 2016, LNCS 9668, pp. 87–103, 2016.
DOI: 10.1007/978-3-319-39636-1_7

sending security guards to visit a large number of sites multiple times over the course of a day in order to fulfill their custodial duty.

Security companies have to schedule tours for their employees in order to cover all needed visits of all objects under their guardianship. The complexity of this task leaves a high potential for solving this problem by algorithmic techniques to minimize the number of needed tours. Thus, we propose the *Districting and Routing Problem for Security Control* (DRPSC) which consists of a districting part and a routing part. In the districting part all objects should be partitioned into a minimum number of disjoint districts, such that a single district can be serviced by a single security guard within each working day of a planning horizon. Given such a partitioning a routing problem has to be solved for each combination of district and day. We seek for a tour starting and ending at a central location which satisfies a maximum tour duration and the time window constraints for each requested visit. In case multiple visits are required at an object in the same period, there typically has to be a separation time between consecutive visits to ensure a better distribution over time. To minimize the number of districts, it is important to minimize the duration of the planned tours in order to incorporate as many objects into the resulting districts as possible, which shows the inseparability of the districting and routing parts.

We address the routing part by an exact *mixed integer linear programming formulation* (MIP) and a *routing construction heuristic* (RCH) with a subsequent *variable neighborhood descent* (VND). For the districting part we propose an *iterative destroy & recreate* (IDR) approach based on an initial solution identified by a *districting construction heuristic* (DCH).

This article is structured as follows. In Sect. 2 a formal problem definition of the DRPSC is given followed by a survey of related work in the literature in Sect. 3. The proposed algorithms for solving the routing subproblem and the districting problem are described in Sects. 4 and 5, respectively. Computational results are shown and discussed in Sect. 6, before final conclusions are drawn and an outlook for possible future work is given in Sect. 7.

2   Problem Definition

This section formalizes the DRPSC. We are given a set of objects $I = \{1, \ldots, n\}$ and a starting location, which we call in relation to the usual terminology in vehicle routing depot 0. There are p planning periods (days) $P = \{1, \ldots, p\}$, and for each object $i \in I$ a set of visits $S_i = \{i_1, \ldots, i_{|S_i|}\}$ is defined. Not all visits, however, have to take place in each period. The visits requested in period $j \in P$ for object $i \in I$ are given by subset $W_{i,j} \subseteq S_i$.

For each visit $i_k \in S_i, i \in I, k = 1, \ldots, |S_i|$, we are given its duration $t_{i_k}^{\text{visit}} \geq 0$ and a time window $T_{i_k} = [T_{i_k}^{\text{e}}, T_{i_k}^{\text{l}}]$, during which the whole visit has to take place. The time windows of successive visits of an object may also overlap but visit i_k always has to take place before a visit $i_{k'}$ with $k, k' \in W_{i,j}$, $k < k'$ and they must be separated by a minimum duration of t^{sep}. The maximum duration of each planned tour must not exceed a global maximum duration t^{max}.

Next, we define underlying graphs on which our proposed algorithms operate. For each period $j \in P$ we define a directed graph $G^j = (V^j, A^j)$ where V^j refers to the set of visits requested at corresponding objects, i.e., $V^j = \bigcup_{i \in I} W_{i,j}$, and the arc set is: $A^j = \{(i_k, i'_{k'}) \mid i_k \in W_{i,j}, i'_{k'} \in W_{i',j}\} \setminus \{(i_k, i_{k'}) \mid i_k, i_{k'} \in W_{i,j}, k' \leq k\}$. We have arc weights associated with every arc in A^j which are given by $t^{\text{travel}}_{i,i'}$, the duration of the fastest connection from object i to object i'. We assume that the triangle inequality holds among these travel times. Let us further define the special nodes 0_0 and 0_1 representing the start and end of a tour and the augmented node set $\hat{V}^j = V^j \cup \{0_0, 0_1\}$, $\forall j \in P$. Accordingly, we add outgoing arcs from node 0_0 to all visits $i_k \in V^j$ and arcs from all visits $i_k \in V^j$ to node 0_1, formally, $\hat{A}^j = A^j \cup \{(0_0, i_k) \mid i_k \in V^j\} \cup \{(i_k, 0_1) \mid i_k \in V^j\}$. Consequently, we define the augmented graph $\hat{G}^j = (\hat{V}^j, \hat{A}^j)$.

The goal of the *DRPSC* is to assign all objects in I to a smallest possible set of districts $R = \{1, \ldots, \delta\}$, i.e., to partition I into δ disjoint subsets I_r, $r \in R$, with $I_r \cap I_{r'} = \emptyset$ for $r, r' \in R$, $r \neq r'$ and $\bigcup_{r \in R} I_r = I$, so that a feasible tour $\tau_{r,j}$ exists for each district I_r, $r \in R$ and each planning period $j \in P$. A tour $\tau_{r,j} = (\tau_{r,j,0}, \tau_{r,j,1}, \ldots, \tau_{r,j,l_{r,j}}, \tau_{r,j,l_{r,j}+1})$ with $\tau_{r,j,0} = 0_0, \tau_{r,j,l_{r,j}+1} = 0_1$, $l_{r,j} = \sum_{i \in I_r} |W_{i,j}|$, and $\tau_{r,j,1}, \ldots, \tau_{r,j,l_{r,j}} \in \bigcup_{i \in I_r} W_{i,j}$ has to start at the depot node 0_0, has to perform each visit $i_k \in W_{i,j}$ in the respective sequence for each object $i \in I_r$ exactly once, and finally has to return back to the depot, i.e., reach node (0_1). A tour $\tau_{r,j}$ is feasible if each visit $\tau_{r,j,u}$, $u = 0, \ldots, l_{r,j} + 1$ takes place in its time window T_{i_k}, where waiting before a visit is allowed, the minimum duration t^{sep} between visits of the same object is fulfilled, and the total tour duration does not exceed $t^{\max}$.

Note that the routing part can be solved for a given district I_r and each period $j \in P$ independently and consists of finding a feasible tour $\tau_{r,j}$.

3   Related Work

To the best of our knowledge there is no work covering all the aspects of the DRPSC as considered here. The similarity of the DRPSC to the vehicle routing problem with time windows (VRPTW), however, leads to a huge amount of related work. A majority of the approaches in the literature aim at minimizing the total route length without taking the makespan into account [4,6,9,11–13,15,16]. As the practical difficulty usually increases when makespan minimization is considered, specialized algorithms have been developed for the traveling salesman problem with time windows (TSPTW) [3,5], which is the specialization of the VRPTW to just one tour. The routing part of the DRPSC is similar to the TSPTW as the aim is to find a feasible tour of duration less than a pre-specified value which is related to the minimization problem of the TSPTW. In the TSPTW, however, multiple visits of the same objects and a separation time between them are not considered. Interestingly, López-Ibáñez et al. [7] showed that by adapting two state-of-the-art metaheuristics for travel time minimization of the TSPTW [6,12] to makespan minimization it is possible to outperform the specialized algorithms. Many of the proposed approaches focus on first minimizing the number of needed routes and only in a second step minimizing the travel time or makespan, e.g., by using a hierarchical objective function [11,13]. Nagata and Bräysy [10] propose a route minimization heuristic which in particular tries to minimize the number of routes needed to solve the VRPTW. They also rely on a destroy-and-recreate heuristic which iteratively tries to delete one route while maintaining an ejection pool (EP). The EP stores all objects which are yet to be inserted. The algorithm tries to identify objects which are difficult to insert in one of the current routes and utilizes this information for choosing objects to be removed and re-inserted. As this approach produced excellent results we adopt this basic idea of the destroy-and-recreate heuristic here.

Exact solution approaches for the VRPTW were proposed by Ascheuer et al. [1] who developed a branch-and-cut algorithm using several valid inequalities and were able to solve most instances with up to 50–70 nodes. Baldacci et al. [2] introduce the ngL-tour relaxation. By using column generation as well as dynamic programming they are able to solve instances with up to 233 nodes to optimality and report new optimal solutions that have not been found previously. A current state-of-the-art method for heuristically solving several variants of the VRPTW is a hybrid genetic algorithm (GA) by Vidal et al. [16]. As many other approaches described in the literature [11,13] they use a penalty function for handling infeasible routes, which is described in [11]. In the GA the initial solutions are created randomly but there are also more elaborate construction heuristics available: Solomon [15] proposes several algorithms for constructing only feasible solutions by extending the well-known savings heuristic, a nearest neighbor heuristic, and insertion heuristics using different criteria. Numerous simple construction heuristics for the asymmetric TSPTW are also proposed by Ascheuer et al. [1].

4   Routing Problem

An important factor when approaching the DRPSC is a practically efficient approach to the underlying routing problems. This part is embedded in the whole approach for optimizing the districting as a subcomponent which is called when the feasibility of a district needs to be checked. As already mentioned, this subproblem is similar to the well-known TSPTW which has been exhaustively studied in the literature. There is, however, one substantial and significant difference: objects have to be visited several times per period and between every two visits of the same object there has to be a specific separation time. Nevertheless, many fruitful ideas of the literature can be adopted to our problem.

As a single routing problem is solved for each period $j \in P$ and each district $r \in R$ independently, we are given one graph $G_r^j = (V_r^j, A_r^j)$. The node set is defined as $V_r^j = V^j \cap \bigcup_{i \in I_r} W_{i,j}$ and the arc set as $A_r^j = A^j \setminus \{(i_k, i'_{k'}) \mid i_k \notin V_r^j \vee i'_{k'} \notin V_r^j\}$. Similarly, we define the augmented graph containing the tours' start and end nodes 0_0 and 0_1 as $\hat{G}_r^j = (\hat{V}_r^j, \hat{A}_r^j)$ where $\hat{V}_r^j = \hat{V}^j \cap \bigcup_{i \in I_r} W_{i,j}$ and $\hat{A}_r^j = \hat{A}^j \setminus \{(i_k, i'_{k'}) \mid i_k \notin \hat{V}_r^j \vee i'_{k'} \notin \hat{V}_r^j\}$.

For computing the duration of a tour τ we first define the arrival and waiting times for each visit of the tour. Moreover, let us define the auxiliary function $\kappa : V_r^j \mapsto I$ which maps the visit $i_k \in V_r^j$, to its corresponding object $i \in$

I_r, and the auxiliary function $\gamma : V_r^j \mapsto \mathbb{N}$ which maps visit $i_k \in V_r^j$, to its corresponding index in the set of visits for this particular object. For every visit $i_k \in V_r^j$, a_{i_k} denotes the arrival time at the object, whereas a_{0_0} and a_{0_1} denote the departure and arrival time for the depot nodes 0_0 and 0_1, respectively. Let $t_{\tau_u}^{\text{wait}} = \max(0, T_{\tau_u}^e - \max(a_{\tau_{u-1}} + t_{\tau_{u-1}}^{\text{visit}} + t_{\kappa(\tau_{u-1}),\kappa(\tau_u)}^{\text{travel}}, a_{\kappa(\tau_u)_{\gamma(\tau_u)-1}} + t_{\kappa(\tau_u)_{\gamma(\tau_u)-1}}^{\text{visit}} + t^{\text{sep}}))$ denote the waiting time before a visit τ_u can be fulfilled. We aim at finding a feasible tour $\tau = (0_0, \tau_1, \ldots, \tau_l, 0_1)$, $\tau_1, \ldots, \tau_l \in V_r^j, l = |V_r^j|$ through all visits starting and ending at the depot such that the total tour duration $T(\tau) = a_{0_1} - a_{0_0}$ does not exceed $t^{\max}$.

4.1   Exact Mixed Integer Linear Programming Model

The following compact mixed integer programming (MIP) model operates on the previously defined and reduced graph G_r^j and is based on Miller-Tucker-Zemlin (MTZ) [8] constraints. We use binary decision variables $y_{i_k,i_{k'}'}, \forall (i_k, i_{k'}') \in A_r^j$ which are set to 1 if the arc between the k-th visit of object i and the k'-th visit of object i' is used in the solution, and 0 otherwise. We model arrival times by additional continuous variables $a_{i_k} \forall i_k \in V_r^j$ and ensure by these variables compliance with the time windows and the elimination of subtours. For each district $r \in R$ and each period $j \in P$ we solve the following model:

$$\min \quad \sum_{i_k \in V_r^j} (t_{i_k}^{\text{wait}} + t_{i_k}^{\text{visit}}) + \sum_{(i_k, i_{k'}') \in \hat{A}_r^j} (y_{i_k,i_{k'}'} \cdot t_{\kappa(i_k),\kappa(i_{k'}')}^{\text{travel}}) \tag{1}$$

$$\text{s.t.} \quad \sum_{(i_k, i_{k'}') \in \hat{A}_r^j} y_{i_k,i_{k'}'} = \sum_{(i_{k'}', i_k) \in \hat{A}_r^j} y_{i_{k'}', i_k} \qquad \forall i_k \in V_r^j \tag{2}$$

$$\sum_{(0_0, i_k) \in \hat{A}_r^j} y_{0_0, i_k} = 1 \tag{3}$$

$$\sum_{(i_k, 0_1) \subset \hat{A}_r^j} y_{i_k, 0_1} = 1 \tag{4}$$

$$a_{i_k} - a_{i_{k'}'} + t^{\max} \cdot (1 - y_{i_{k'}', i_k}) \geq t_{\kappa(i_{k'}'),\kappa(i_k)}^{\text{travel}} + t_{i_{k'}'}^{\text{visit}}$$
$$\forall i_k \in \hat{V}_r^j, (i_k, i_{k'}') \in \hat{A}_r^j \tag{5}$$

$$a_{i_k} + t_{0,\kappa(i_k)}^{\text{travel}} \cdot (1 - y_{0_0, i_k}) \geq t_{0,\kappa(i_k)}^{\text{travel}} \qquad \forall (0_0, i_k) \in \hat{A}_r^j \tag{6}$$

$$t_{i_k}^{\text{wait}} + t^{\max} \cdot (1 - y_{i_k, i_{k'}'}) \geq a_{i_{k'}'} - a_{i_k} - t_{\kappa(i_k),\kappa(i_{k'}')}^{\text{travel}} - t_{i_k}^{\text{visit}}$$
$$\forall i_k \in \hat{V}_r^j, (i_k, i_{k'}') \in \hat{A}_r^j \tag{7}$$

$$a_{i_{k-1}} \leq a_{i_k} - t^{\text{sep}} \qquad \forall i_k, i_{k-1} \in V_r^j \tag{8}$$

$$\sum_{(i_k, i_{k'}') \in \hat{A}_r^j} y_{i_k, i_{k'}'} = 1 \qquad \forall i_k \in V_r^j \tag{9}$$

$$T_{i_k}^e \leq a_{i_k} \leq T_{i_k}^l - t_{i_k}^{\text{visit}} \qquad \forall i_k \in V_r^j \tag{10}$$

$$y_{i_k, i_{k'}'} \in \{0,1\} \qquad \forall (i_k, i_{k'}') \in \hat{A}_r^j \tag{11}$$

The objective function (1) minimizes the total makespan within which all object visits take place by summing up all visit times, travel times, and waiting times. Equalities (2) ensure that the number of ingoing arcs is equal to the number of outgoing arcs for each node $i_k \in V_r^j$. Equalities (3) and (4) ensure that there must be exactly one ingoing and outgoing arc for the depot in each period $j \in P$. Inequalities (5) are used to recursively compute the arrival times for every visit. If an edge $(i_k, i'_{k'})$ is not used, then the constraint is deactivated. These inequalities can be individually lifted by using $(T_{i'_{k'}}^l - t_{i'_{k'}}^{\text{visit}}) + (t_{\kappa(i'_{k'}),\kappa(i_k)}^{\text{travel}} + t_{i'_{k'}}^{\text{visit}})$ instead of t^{tmax}, which is also done in our implementation. Inequalities (6) set the start time at the depot for each period. Inequalities (7) compute the waiting time at the k-th visit of object i before traveling to the k'-th visit of object i'. We need these waiting times $t_{i_k}^{\text{wait}} \, \forall i_k \in V_r^j$ for the objective function to minimize the makespan of the route. These inequalities can also be lifted by replacing $t^{\max}$ with the term $(T_{i'_{k'}}^l - t_{i'_{k'}}^{\text{visit}}) - T_{i_k}^l - t_{\kappa(i_k),\kappa(i'_{k'})}^{\text{travel}} - t_{i_k}^{\text{visit}}$. Inequalities (8) model the minimum time required between two different visits of the same object, i.e., ensure the separation time t^{sep}. Inequalities (9) state that there must exist an ingoing and an outgoing arc for the k-th visit of object i, if this particular visit is requested in the considered period $j \in P$. It is ensured that every time window of every visit $i_k \in V_r^j$ is fulfilled in (10). In (11) the domain definitions for the binary edge-decision variables $y_{i_k,i'_{k'}}$ are given.

In the context of the districting problem we use this model only for checking feasibility which can usually be done faster than solving the optimization problem to optimality. To this end we replace the objective function by $\min\{0\}$ and add the following constraints for limiting the makespan to $t^{\max}$:

$$\sum_{i_k \in V_r^j} (t_{i_k}^{\text{wait}} + t_{i_k}^{\text{visit}}) + \sum_{(i_k,i'_{k'}) \in \hat{A}_r^j} (y_{i_k,i'_{k'}} \cdot t_{\kappa(i_k),\kappa(i'_{k'})}^{\text{travel}}) \leq t^{\max} \qquad (12)$$

4.2 Heuristics

For larger districts the exact feasibility check using the MIP model might be too slow, hence we also propose a faster greedy construction heuristic followed by a variable neighborhood descent.

Given a sequence of visits τ, we first determine if a tour can be scheduled such that the time window constraints of all visits are satisfied. For this purpose, we compute the earliest possible arrival time a_{i_k} for each visit and minimize waiting times.

Feasibility of a Tour: Since the (intermediate) tour τ starts at the depot at the earliest possible time, the departure at the depot a_{0_0} is set to 0. For each subsequent visit τ_u, the arrival time a_{τ_u} is the maximum of $T_{\tau_u}^e$ and the arrival time at the preceding visit $a_{\tau_{u-1}}$ including visit time $t_{\tau_{u-1}}^{\text{visit}}$ and travel time $t_{\kappa(\tau_{u-1}),\kappa(\tau_u)}^{\text{travel}}$ from the preceding visit's object $\kappa(\tau_{u-1})$ to the current visit's object $\kappa(\tau_u)$. The depot has no requested visit times, therefore we define $t_{0_0}^{\text{visit}} =$

$t_{0_1}^{\text{visit}} = 0$. Furthermore, for each object i the separation time t^{sep} between visit i_k and i_{k-1} for all $k > 1$ has to be respected. Formally:

$$a_{0_0} = 0$$

$$a_{\tau_u} = \begin{cases} \max\{T_{\tau_u}^e, a_{\tau_{u-1}} + t_{\tau_{u-1}}^{\text{visit}} + t_{\kappa(\tau_{u-1}),\kappa(\tau_u)}^{\text{travel}}\} \\ \qquad\qquad\qquad\qquad\qquad \text{for } u > 1, \gamma(\tau_u) = 1 \\ \max\{T_{\tau_u}^e, a_{\tau_{u-1}} + t_{\tau_{u-1}}^{\text{visit}} + t_{\kappa(\tau_{u-1}),\kappa(\tau_u)}^{\text{travel}}, \\ \qquad a_{\kappa(\tau_u)_{\gamma(\tau_u)-1}} + t_{\kappa(\tau_u)_{\gamma(\tau_u)-1}}^{\text{visit}} + t^{\text{sep}}\} \\ \qquad\qquad\qquad\qquad\qquad \text{for } u > 1, \gamma(\tau_u) > 1 \end{cases}$$

$$a_{0_1} = a_{\tau_l} + t_{\tau_l}^{\text{visit}} + t_{\kappa(\tau_l),0}^{\text{travel}}$$

If for any arrival time a_{i_k} with $i_k \in V_r^j$ the following condition is violated, the sequence of visits is infeasible:

$$a_{i_k} + t_{i_k}^{\text{visit}} \leq T_{i_k}^l \tag{13}$$

The resulting tour duration $T(\tau) = a_{0_1} - a_{0_0}$ can be minimized while keeping τ feasible by delaying the departure at the depot by the so called *forward time slack* proposed by Savelsbergh [14] for the TSPTW. The *forward time slack* $F(\tau_u, \tau_{u'})$ for the partial tour $\tau' = \tau_u, \ldots, \tau_{u'}$ adapted to our problem is

$$F'(\tau_u, \tau_{u'}) = \begin{cases} T_{\tau_{u'}}^l - t_{\tau_{u'}}^{\text{visit}} - a_{\tau_u} \\ \qquad\qquad\qquad\qquad\qquad \text{for } u = u' \\ F'(\tau_u, \tau_{u'-1}) - T_{\tau_{u'-1}}^l + T_{\tau_{u'}}^l - t_{\tau_{u'}}^{\text{visit}} - t_{\kappa(\tau_{u'-1}),\kappa(\tau_{u'})}^{\text{travel}} \\ \qquad\qquad\qquad\qquad\qquad \text{for } u' > 1, \gamma(\tau_{u'}) = 1 \\ \min\{F'(\tau_u, \tau_{u'-1}) - T_{\tau_{u'-1}}^l + T_{\tau_{u'}}^l - t_{\tau_{u'}}^{\text{visit}} - t_{\kappa(\tau_{u'-1}),\kappa(\tau_{u'})}^{\text{travel}}, \\ \qquad F'(\tau_u, \tau_{\kappa(\tau_{u'})_{\gamma(\tau_{u'})-1}}) - T_{\tau_{\kappa(\tau_{u'})_{\gamma(\tau_{u'})-1}}}^l + T_{u'}^l - t_{\tau_{u'}}^{\text{visit}} - t^{\text{sep}}\} \\ \qquad\qquad\qquad\qquad\qquad \text{for } u' > 1, \gamma(\tau_{u'}) > 1 \end{cases}$$

$$F(\tau_u, \tau_{u'}) = \min_{v=u,\ldots,u'} \{F'(\tau_u, \tau_v)\} \tag{14}$$

The tour duration of tour τ must not exceed t^{max}. Formally, if

$$T(\tau) - \min(F(0_0, 0_1), \sum_{u=1}^{l} t_{\tau_u}^{\text{wait}}) < t^{\text{max}} \tag{15}$$

holds, the sequence τ is feasible, otherwise infeasible.

Routing Construction Heuristic: We developed a *Routing Construction Heuristic* (RCH) based on an insertion heuristic by starting from a partial tour $\tau' = (0_0, 0_1)$ containing only the start and end nodes and iteratively adding all visits $i_k \in V_r^j$ to τ'. A 2-step approach is used, where we first order the visits according to some criteria and then insert them at the first feasible or best possible insert position, respectively. For the insertion order we compute the *flexibility value* of each visit $i_k \in V_r^j$ where visits with less flexibility are inserted first:

$$flex(i_k) = T_{i_k}^l - T_{i_k}^e - t_{i_k}^{\text{visit}} \tag{16}$$

$$flex(i_k^{(1)}) \leq flex(i_k^{(2)}) \leq \cdots \leq flex(i_k^{(|V_r^j|)}) \tag{17}$$

Visits with less flexibility may be more difficult to insert as they need to be scheduled at a very specific time. Ties are broken randomly.

In a second phase we start by trying to insert the first visit, i.e., $i_k^{(1)}$, into the partial tour τ'. We start at the front, i.e., try to insert it after the start node 0_0, and move backwards to the end. Then, we either stop when we found the first feasible insert position in the first feasible variant or we compute insertion costs for each possible insert position and insert the visit at the position with the minimum costs for the best possible insertion variant. We define these costs as:

$$d_{i_k,u'} = \begin{cases} a_{\tau_{u'}} + t_{\tau_{u'}}^{\text{visit}} + t_{\kappa(\tau_{u'}),\kappa(\tau_u)}^{\text{travel}} - a_{\tau_u} & \text{if (19) and (20) hold} \\ \infty & \text{otherwise} \end{cases} \tag{18}$$

These insertion costs $d_{i_k,u'}$ determine the amount of time by which the visit τ_u has to be moved backwards in order to insert the new visit $\tau_{u'}$. Note that $d_{i_k,u'}$ may also be negative, if the space for insertion of visit $\tau_{u'}$ is bigger than necessary. However, this is desirable as we use those insert positions more likely which have bigger gaps and smaller gaps are kept for later inserts. In Sect. 6 we compare both, the first feasible and the best possible variant, to each other in terms of solution quality and runtime.

We further maintain global variables for the *forward time slack* $F(\tau')$ and all arrival times a_{τ_u} of each partial tour τ' computed during the execution of the insertion heuristic. For an insertion to be feasible, the latest allowed arrival time at visit $\tau_{u'}$ must be greater or equal to the earliest possible arrival at that visit considering the previous visit's earliest arrival, its visit time and the travel time between τ_{u-1} and $\tau_{u'}$. Furthermore, the earliest departure at $\tau_{u'}$ including the travel time between $\tau_{u'}$ and τ_u must be smaller or equal to the earliest arrival at τ_u delayed by the *forward time slack* of the partial tour from τ_u to the depot. Using the definition of the forward time slack in equality (14) and if inequality (13) holds, then the insertion is feasible if, in addition, also the following two inequalities hold:

$$T_{\tau_{u'}}^l - t_{\tau_{u'}}^{\text{visit}} \geq \max\{a_{\tau_{u-1}} + t_{\tau_{u-1}}^{\text{visit}} + t_{\kappa(\tau_{u-1}),\kappa(\tau_{u'})}^{\text{travel}}, a_{\kappa(\tau_u)_{\gamma(\tau_u)-1}} + t_{\kappa(\tau_u)_{\gamma(\tau_u)-1}}^{\text{visit}} + t^{\text{sep}}\} \tag{19}$$

$$a_{\tau_{u'}} + t_{\tau_{u'}}^{\text{visit}} + t_{\kappa(\tau_{u'}),\kappa(\tau_u)}^{\text{travel}} \leq a_{\tau_u} + F(\tau_u, 0_1) \tag{20}$$

Local Improvement: If the solution found by the RCH is infeasible we additionally employ a VND to reduce the number of infeasibilities and possibly come to a feasible solution. First, we insert each infeasible visit i_k into the tour on the position u' where the costs $d_{i_k,u'}$ are minimum. We use a lexicographical penalty function to penalize infeasible tours where the first criterion is the number of time window violations and the second criterion is the duration of the route as proposed by López-Ibáñez et al. [6]. We use three common neighborhood structures from the literature and search them in a best improvement fashion in random order while respecting the visit order:

Swap: This neighborhood considers all exchanges between two distinct visits.
2-opt: This is the classical 2-opt neighborhood for the traveling salesman problem where all edge exchanges are checked for improvement.
Or-opt: This neighborhood considers all solutions in which sequences of up to three consecutive visits are moved to another place in the same route.

If at some point during the algorithm the value of the penalty function is zero we terminate with a feasible solution.

5   Districting Problem

In the previous section we have already introduced a fast heuristic for efficiently testing feasibility of a given set of objects by building a single tour for each period through all requested visits of these objects. In the districting part of the DRPSC we face the problem of intelligently assigning objects to districts such that the number of districts is minimized. For checking the feasibility of this assignment we use the previously introduced RCH. Alternatively, we could also use our MIP model for solving these subproblems but, as we will see in Sect. 6, it is too slow to be used in practical scenarios. We propose a DCH and an iterative destroy & recreate algorithm where the former generates an initial solution and the latter tries to iteratively remove districts.

5.1   Districting Construction Heuristic

Starting with one district, objects are iteratively added to the existing districts $r \in R$. Whenever adding an object $i \in U$ to any of the available districts $r \in R$ would make the assignment infeasible, $i \in U$ is added to a newly created district r'. The overall DCH is shown in Algorithm 1 and explained below.

First, the set of districts R is initialized with the first empty district 1 and the set of objects I is sorted by extending the flexibility values as defined in Eq. (16) from visits to objects. All objects are sorted by the sum of their flexibility values $\sum_{j \in P} \sum_{i_k \in W_{i,j}} flex(i_k)$ in ascending order. As in the RCH, the resulting set U is denoted as the set of unscheduled visits. The DCH terminates when all $i \in U$ have been scheduled (2) and, as a consequence, all requested visits have been inserted successfully and we obtain a feasible solution to the DRPSC. The insertion of object i into district r (lines 5 and 13) is accomplished by checking

---

Algorithm 1. Districting Construction Heuristic

1: **init:** $R \leftarrow \{1\}, U \leftarrow sort(I)$
2: **for all** $i \in U$ **do**
3:     $inserted \leftarrow$ **false**
4:     **for all** $r \in R$ **do**
5:         **if** $insert(i, r)$ **then**
6:             $inserted \leftarrow$ **true**
7:             $break$
8:         **end if**
9:     **end for**
10:    **if not** $inserted$ **then**
11:        $r' \leftarrow$ create $|R| + 1$-th new empty district
12:        $R \leftarrow R \cup \{r'\}$
13:        $insert(i, r')$
14:    **end if**
15: **end for**

---

for each scheduled visit $i_k \in W_{i,j}$ if i_k can be feasibly inserted into the particular district r (for the definition of feasibility of a tour see also Sect. 4.2). In line 5 the DCH inserts i either into the first feasible or into the best possible insert position, as described in Sect. 4.2. The *insert* function returns *false*, if no feasible insertion position is found for at least one $i_k \in W_{i,j}$, $\forall j \in P$. It returns *true*, if a feasible insertion position is found for each visit $i_k \in W_{i,j}$, $\forall j \in P$. If the loop over all districts (line 4) terminates without finding any feasible insertion position the variable *inserted* stays *false* and a new empty district is created in line 11. The proposed constructive algorithm will terminate with a feasible solution after $|U|$ iterations.

5.2 Iterative Destroy and Recreate

Nagata and Bräysy [10] proposed a *route elimination algorithm* for reducing the number of vehicles needed in the VRPTW. We apply the basic idea to the districting problem. The algorithm starts with the initial assignment where every object is reached by a separate route. Then, one district $r \in R$ is chosen for elimination at a time, maintaining all now unassigned objects in an *ejection pool* (EP). Then, it is tried to assign all objects of the EP to the remaining districts $R \setminus \{r\}$. If this is successful, the number of districts could be reduced by one and another district is chosen for elimination. We adapt this idea to the DRPSC and use the result of the DCH described in Sect. 5.1 as initial solution.

Let the assignment of an object $i \in I$ to a district $r \in R$ be feasible if and only if a feasible tour can be scheduled for all assigned visits of all objects for each period. Let c_i be a penalty value of object $i \in I$ denoting failed attempts of inserting object i into a district. Each time a visit cannot be inserted, this penalty value is increased by one, revealing objects which are difficult to assign to one of the available districts.

Algorithm 2. District elimination algorithm

```
 1: init: EP ← ∅, c_i ← 0 ∀i ∈ I
 2: choose a district r^del ∈ R for deletion
 3: R ← R \ {r^del}
 4: EP ← EP ∪ {i | i ∈ I_{r^del}}
 5: while EP ≠ ∅ ∧ termination criterion not met do
 6:    i^ins ← arg max_{i∈EP} {c_i}
 7:    R_f ← feasible districts for assignment of i^ins
 8:    c_{i^ins} ← c_{i^ins} + |R| − |R_f|
 9:    if R_f ≠ ∅ then
10:       assign object i^ins to a randomly chosen feasible district r ∈ R_f
11:    else
12:       select random district r^ins ∈ R
13:       assign object i^ins to district r^ins
14:       call VND for district r^ins (see Sect. 4.2)
15:       while ∃ an infeasible tour for any period of district r^ins do
16:          i^del ← arg min_{i∈I_{r^ins}} {c_i}
17:          I_{r^ins} ← I_{r^ins} \ {i^del}
18:          EP ← EP ∪ {i^del}
19:          call VND for district r^ins (see Sect. 4.2)
20:       end while
21:    end if
22: end while
```

If the EP becomes empty, a feasible assignment of objects to districts is found. Subsequently, another iteration is started, destroying a district and reassigning its objects to the remaining ones. The overall district elimination algorithm is shown in Algorithm 2.

First, the EP is initialized to the empty set and the penalty values of all objects are set to 0. Starting with the solution provided by DCH a district is chosen for elimination in line 2. One of the following strategies is applied uniformly at random for selecting a district for elimination:

Minimum number of scheduled visits: This implies that only a minimum number of visits has to be reinserted to regain a feasible solution.

Shortest tour duration: Selecting a district where the maximum tour duration over all periods is minimal can be promising because this district might lead to a district with visits of shorter durations resulting in easier insert operations.

Maximum waiting times: Selecting a district with a loose schedule may indicate less or shorter visits, making them easier to reinsert.

After deleting a district all objects of this district are moved to the EP (line 4). As long as the EP contains objects, we try to assign each object to one of the remaining districts. An object with maximum penalty value is chosen for the next assignment (5). For the chosen object i^{ins} the feasible districts for assignment are computed. If there is at least one feasible district for an assignment of object i^{ins} (9) we assign the object to such a district uniformly at random (10). If it is not possible to feasibly assign the object i^{ins} to any of the remaining districts we randomly choose a district for assignment (11). Then, we apply the VND described in Sect. 4.2 trying to make the district feasible. If this is not possible and the assignment is still infeasible we iteratively try to remove objects with lowest penalty values from this district r^{ins} in the following loop (15), remove them from district r^{ins} (17), and finally add them to the EP (18). Then again, we call the VND from Sect. 4.2 trying to make the resulting tour from the actual assignment feasible. After an iteration of the outer loop the object with highest penalty value of the EP has been inserted and other objects previously assigned to this district may have been added to the EP. The idea behind this approach is to insert difficult objects first and temporarily remove easy to insert objects from the solution to reinsert them later. When the EP is empty, a new best assignment with one district less is found. This algorithm iterates until a termination criterion, e.g., a time limit is met.

6   Computational Results

To evaluate our proposed algorithm computational tests are performed on a benchmark set of instances. As the DRPSC is a new problem we created new instances[1] based on the characteristics of real-world data provided by an industry partner. The main characteristics of real-world data are: Most of the time windows are of medium size, the depot is centralized among the objects, travel times are rather small with respect to visit times and the number of visits of the objects is usually ranged from 1 to 4. The distance matrix is taken from TSPlib instances and we added the depot, the visits and the time windows in the following way: The depot is selected by taking the node for which the total distance to all other nodes is a minimum. Each node of the original instance has between 1 and v visits, where v is a parameter of the instance. Small time windows have a length between 5 and 30 min, medium time windows have a length of 2, 3, 4, or 5 h, and visits with large time windows are unrestricted. For the instance generation the length of a time window is assigned randomly to a visit based on parameter values α and β: a small time window is chosen with probability α, a medium time window with probability β, and a large time window with probability $1 - \alpha - \beta$. Furthermore, we enforce that small and medium time windows of visits of the same object do not overlap and we choose the visit time uniformly at random from 3 to 20 min. For all our instances we set t^{sep} to 60 min and t^{max} to 10 h.

The algorithm is implemented in C++ using Gurobi 6.5 for solving the MIP. For each combination of configuration and instance we performed 20 independent runs for the IDR while for the routing part we performed only one run because all tested algorithms for the routing part are deterministic. All runs were executed on a single core of an Intel Xeon processor with 2.54 GHz. The iterative destroy & recreate algorithm is terminated after a maximum of 900 CPU seconds. The MIP model for the routing part was aborted after 3600 CPU seconds.

---

[1] https://www.ac.tuwien.ac.at/files/resources/instances/drpsc/hm16.tar.gz.

In the first set of experiments the routing part of the DRPSC is examined more closely to evaluate RCH in comparison to the MIP model. Then, several configurations of our proposed algorithm for the whole problem are investigated.

6.1 Routing Part

First, the methods for the routing part are evaluated on a separate set of benchmark instances. In Table 1 the MIP model is compared to RCH, and RCH with the subsequent VND, denoted by RCH-VND. As the goal for the routing part is to minimize the makespan of a specific route, the maximum tour duration constraint is relaxed and the resulting makespan is given in minutes in the column obj. In the first four columns the instance parameters are specified. Sequentially, the instance name, the number of objects $|I|$, the maximum number of nodes of all objects $|V|$, the percentage of small (α), and medium time windows (β) and the maximum number of visits per objects v is given. For the RCH and RCH-VND we give the objective value (makespan in minutes) and the time needed for solving the instance. Then, the upper bound (UB), the lower bound (LB), the final optimality gap, and the time spent by Gurobi for solving the MIP model is shown. In the two remaining columns we present the relative gap between the MIP and RCH-VND $\Delta_{\text{MIP}} = (obj_{\text{RCH-VND}} - LB)/obj_{\text{RCH-VND}}$ as well as the relative gap between RCH and RCH-VND $\Delta_{\text{RCH}} = (obj_{\text{RCH}} - obj_{\text{RCH-VND}})/obj_{\text{RCH-VND}}$.

In Table 1 we see that the MIP model is able to solve easier instances to optimality, but soon has very high running times. RCH-VND yields very reasonable solutions with objective values close to the LB of the MIP for most cases. When looking at the relative gap between the RCH and RCH-VND (Δ_{RCH}), we can conclude that the VND improves greatly on the objective value with only a minor increase in running time. Moreover, Δ_{MIP} reveals that RCH-VND produces results close to the results of the MIP, and for those instances where the relative gap between the MIP and RCH-VND is greater than 10 %, the MIP also has a relatively larger gap between UB and LB.

As we require a fast method for deciding if a route is feasible within the districting problem, we conclude that RCH-VND is a reasonable choice.

6.2 Districting Part

For testing the proposed algorithms for the DRPSC we used three different configurations. In the IDR-DCH[f] algorithm we used the DCH for generating an initial feasible solution candidate with the first feasible strategy in contrast to the IDR-DCH[b] where we used a best possible strategy. Both configurations are compared with the IDR-SCH, where a simple construction heuristic (SCH) as proposed by Nagata and Bräysy [11] is used. In the SCH each object is put in a separate district which results in a trivial initial solution candidate.

In Table 2 the results of the experiments are shown. Columns obj show the average objective value, i.e., the minimum number of districts at the end of optimization, after the full run of IDR while columns obj_f and obj_b show the

Table 1. Results of the MIP, RCH, and RCH-VND for the routing part.

Instance						RCH		RCH-VND		MIP				Rel. Difference	
name	$|I|$	$|V|$	α	β	v	obj	t[s]	obj	t[s]	UB	LB	Gap	t[s]	Δ_{MIP}	Δ_{RCH}
burma14_01	13	19	0	0.2	2	495.80	< 0.01	333.49	0.03	332.62	332.62	0.00 %	2025.56	0.26 %	48.67 %
burma14_02	13	20	0	0.2	2	624.47	< 0.01	374.60	0.02	352.93	343.50	2.67 %	3600.00	8.30 %	66.70 %
burma14_03	13	21	0	0.2	2	525.49	0.01	440.86	0.05	433.42	421.91	2.66 %	3600.00	4.30 %	19.20 %
burma14_04	13	19	0	0.2	2	607.11	< 0.01	397.14	0.03	395.15	387.89	1.84 %	3600.00	2.33 %	52.87 %
burma14_05	13	22	0	0.2	2	606.07	< 0.01	409.24	0.03	356.71	337.43	5.40 %	3600.00	17.55 %	48.10 %
burma14_06	13	19	0	0.2	2	409.54	< 0.01	273.28	0.01	272.93	272.93	0.00 %	2318.17	0.13 %	49.86 %
burma14_07	13	23	0	0.5	2	714.04	< 0.01	508.69	0.05	493.48	456.82	7.43 %	3600.00	10.20 %	40.37 %
burma14_08	13	17	0	0.5	2	372.63	< 0.01	312.70	0.02	311.10	311.10	0.00 %	2.09	0.51 %	19.16 %
burma14_09	13	21	0	0.5	2	416.55	< 0.01	370.61	0.04	360.71	360.71	0.00 %	3.35	2.67 %	12.40 %
burma14_10	13	20	0	0.5	2	386.02	< 0.01	342.00	0.02	336.25	336.25	0.00 %	10.74	1.68 %	12.87 %

Table 2. Results of the DCH and IDR for the districting part.

Instance							IDR-SCH			IDR-DCH^f					IDR-DCH^b				
name	runs	$|I|$	$|V|$	c	β	v	obj	sd	t^*[s]	obj_f	t_f[s]	obj	sd	t^*[s]	$\tilde{obj}_b$	t_b	obj	sd	t^*[s]
st70_1	20	69	105	0.1	0.7	2	3.0	0.0	11	6	< 0.1	3.0	0.0	5	15	0.1	3.0	0.0	9
st70_2	20	69	91	0.1	0.7	2	3.0	0.0	4	6	< 0.1	3.0	0.0	6	15	0.1	3.0	0.0	6
st70_3	20	69	106	0.1	0.7	2	3.0	0.0	11	6	< 0.1	3.0	0.0	13	15	0.1	3.0	0.0	11
rd100_1	20	99	152	0.1	0.5	2	6.0	0.0	8	9	< 0.1	6.0	0.0	10	9	0.1	6.0	0.0	14
rd100_2	20	99	160	0.1	0.5	2	6.0	0.0	16	8	0.1	6.0	0.0	14	8	0.1	6.0	0.0	6
rd100_3	20	99	152	0.1	0.5	2	6.0	0.0	3	7	0.1	6.0	0.0	2	8	0.1	6.0	0.0	7
tsp225_1	20	224	334	0.2	0.7	2	11.0	0.0	47	13	0.2	11.0	0.0	64	15	0.4	11.0	0.0	121
tsp225_2	20	224	341	0.2	0.7	2	11.0	0.0	46	13	0.3	11.0	0.0	67	15	0.5	11.0	0.0	36
tsp225_3	20	224	332	0.2	0.7	2	10.0	0.0	226	12	0.3	10.0	0.0	257	12	0.5	10.0	0.0	177
gr48_1	20	47	120	0.2	0.7	4	5.0	0.0	6	8	< 0.1	5.0	0.0	4	7	< 0.1	5.0	0.0	14
gr48_2	20	47	115	0.2	0.7	4	5.0	0.0	12	7	< 0.1	5.0	0.0	7	7	< 0.1	5.0	0.0	10
gr48_3	20	47	125	0.2	0.7	4	4.0	0.0	527	7	< 0.1	4.0	0.0	438	6	< 0.1	4.0	0.0	488
berlin52_1	20	51	133	0.0	0.7	4	5.0	0.0	3	6	< 0.1	5.0	0.0	2	7	0.1	5.0	0.0	7
berlin52_2	20	51	130	0.0	0.7	4	5.0	0.0	5	7	< 0.1	4.0	0.0	142	6	0.1	5.0	0.0	1
berlin52_3	20	51	140	0.0	0.7	4	5.0	0.0	19	7	< 0.1	5.0	0.0	14	6	0.1	5.0	0.0	16
ft70_1	20	69	167	0.1	0.5	4	8.0	0.0	12	12	< 0.1	8.0	0.0	12	10	0.1	8.0	0.0	18
ft70_2	20	69	180	0.1	0.5	4	8.0	0.0	33	11	< 0.1	8.0	0.0	15	11	0.1	8.0	0.0	18
ft70_3	20	69	144	0.1	0.5	4	7.0	0.0	41	9	< 0.1	7.0	0.0	36	9	0.1	7.0	0.0	19
ch150_1	20	149	360	0.2	0.5	4	11.0	0.0	539	15	0.2	11.0	0.0	480	15	0.4	11.2	0.4	881
ch150_2	20	149	402	0.2	0.5	4	12.0	0.0	342	17	0.2	12.0	0.0	338	15	0.4	12.0	0.0	387
ch150_3	20	149	357	0.2	0.5	4	11.0	0.0	655	14	0.2	11.0	0.0	520	13	0.4	11.0	0.0	808

average objective value, i.e., the average number of districts, after the respective construction heuristic. Columns t^* show the median time in seconds after which the best solution has been found during the run of IDR while t_f and obj_b show the median time after which the respective construction heuristic has found an initial solution. Columns sd show the standard deviation of the objective value for 20 runs of a single instance.

We observe that for most instances the final objective value of the IDR is the same for all three configurations. There are, however, differences for the construction heuristics alone and DCH^b for most but not all instances better results but needed more time. The IDR-SCH works surprisingly well and was able to find good results in about the same amount of time as the other two (more sophisticated) configurations.

7 Conclusions and Future Work

In this work we introduced a new vehicle routing problem which originates from the security control sector. The goal of the Districting and Routing Problem for Security Control is to partition a set of objects under surveillance into disjoint clusters such that for each period a route through all requested visits can be scheduled satisfying complex time window constraints. As the objects may require multiple visits, there needs to be a minimum separation time between each two visits which imposes an interesting additional challenge. The proposed heuristic solution approach starts with a greedy construction heuristic followed by an iterate destroy and recreate algorithm. The latter works by iteratively destroying districts and trying to insert the resulting unassigned objects into the other districts. The computational results reveal that the MIP model is able to solve smaller instances of the routing problem to optimality and that the quality of the initial solutions of the districting problem has only a minor influence on the final solution quality. There are several possibilities for extending this algorithm in future work. As the feasibility check for a district is time-consuming a caching mechanism to prevent checking the same assignment of objects all over again seems promising. This could even be extended to checking subsets of such assignments, which also must be feasible if any superset of these objects results in feasible routes. Another idea is to use neighborhood structures which exchange objects of two or more distinct clusters.

References

1. Ascheuer, N., Fischetti, M., Grötschel, M.: Solving the asymmetric travelling salesman problem with time windows by branch-and-cut. Math. Program. **90**(3), 475–506 (2001)
2. Baldacci, R., Mingozzi, A., Roberti, R.: New state-space relaxations for solving the traveling salesman problem with time windows. INFORMS J. Comput. **24**(3), 356–371 (2012)

3. Cheng, C.B., Mao, C.P.: A modified ant colony system for solving the travelling salesman problem with time windows. Math. Comput. Model. **46**(9), 1225–1235 (2007)

4. Da Silva, R.F., Urrutia, S.: A general VNS heuristic for the traveling salesman problem with time windows. Discrete Optim. **7**(4), 203–211 (2010)

5. Gambardella, L.M., Taillard, E., Agazzi, G.: MACS-VRPTW: A multiple ant colony system for vehicle routing problems with time windows. In: Corne, D., Dorigo, M., Glover, F., Dasgupta, D., Moscato, P., Poli, R., Price, K.V. (eds.) New Ideas in Optimization, Chap. 5, pp. 63–76. McGraw-Hill Ltd., Maidenhead (1999)

6. López-Ibáñez, M., Blum, C.: Beam-ACO for the travelling salesman problem with time windows. Comput. Oper. Res. **37**(9), 1570–1583 (2010)

7. López-Ibáñez, M., Blum, C., Ohlmann, J.W., Thomas, B.W.: The travelling salesman problem with time windows: Adapting algorithms from travel-time to makespan optimization. Appl. Soft Comput. **13**(9), 3806–3815 (2013)

8. Miller, C.E., Tucker, A.W., Zemlin, R.A.: Integer programming formulation of traveling salesman problems. J. ACM **7**(4), 326–329 (1960)

9. Mladenović, N., Todosijević, R., Urošević, D.: An efficient GVNS for solving traveling salesman problem with time windows. Electron. Notes Discrete Math. **39**, 83–90 (2012)

10. Nagata, Y., Bräysy, O.: A powerful route minimization heuristic for the vehicle routing problem with time windows. Oper. Res. Lett. **37**(5), 333–338 (2009)

11. Nagata, Y., Bräysy, O., Dullaert, W.: A penalty-based edge assembly memetic algorithm for the vehicle routing problem with time windows. Comput. Oper. Res. **37**(4), 724–737 (2010)

12. Ohlmann, J.W., Thomas, B.W.: A compressed-annealing heuristic for the traveling salesman problem with time windows. INFORMS J. Comput. **19**(1), 80–90 (2007)

13. Prescott-Gagnon, E., Desaulniers, G., Rousseau, L.M.: A branch-and-price-based large neighborhood search algorithm for the vehicle routing problem with time windows. Networks **54**(4), 190–204 (2009)

14. Savelsbergh, M.W.P.: The vehicle routing problem with time windows: Minimizing route duration. ORSA J. Comput. **4**(2), 146–154 (1992)

15. Solomon, M.M.: Algorithms for the vehicle routing and scheduling problems with time window constraints. Oper. Res. **35**(2), 254–265 (1987)

16. Vidal, T., Crainic, T.G., Gendreau, M., Prins, C.: A hybrid genetic algorithm with adaptive diversity management for a large class of vehicle routing problems with time-windows. Comput. Oper. Res. **40**(1), 475–489 (2013)

A GRASP/VND Heuristic for a Generalized Ring Star Problem

Rodrigo Recoba, Franco Robledo, Pablo Romero$^{(\boxtimes)}$, and Omar Viera

Facultad de Ingeniería, Universidad de la República,
Julio Herrera Y Reissig 565, Montevideo, Uruguay
{recoba,frobledo,promero,viera}@fing.edu.uy
http://www.fing.edu.uy

Abstract. A generalization of the Ring Star Problem, called Two-Node-connected Star Problem (2NCSP), is here addressed. We are given an undirected graph, pendant-costs and core-costs. The goal is to find the minimum-cost spanning graph, where the core is a two-node-connected component, and the remaining nodes are pendant to this component.

First, we show that the 2NCSP belongs to the $\mathcal{NP}$-Hard class. Therefore, a GRASP heuristic is developed, enriched with a Variable Neighborhood Descent (VND). The neighborhood structures include exact integer linear programming models to find best paths as well as a shaking operation in order not to get stuck in a local minima.

We contrast our GRASP/VND methodology with prior works from RSP using TSPLIB, in order to highlight the effectiveness of our heuristic. Our solution outperforms several instances considered in a previous reference related to the RSP. A discussion of the results and trends for future work is provided.

Keywords: GRASP · VND · Telecommunications · Ring star problem

1 Introduction

Historically, network design considers two-node-connected topologies, in order to preserve connectivity under single point of failures. A cornerstone in topological network design is credited to Clyde Monma et al. [6]. The authors provide a structural characterization of an optimum solution. Surprisingly, the best ring is not always the optimal 2-connected topology.

Inspired in fiber optics design, M. Labbé et al. introduce the Ring Star Problem, or RSP for short [5]. It can be understood as a generalization of the Hamiltonian Tour, where some terminals may be pendant from the ring (i.e., they are directly connected by some node that belongs to the ring). The hardness of the RSP follows from the fact that Hamiltonian Tour is a hard combinatorial problem. Therefore, in practice the RSP has been addressed heuristically, in several oportunities, and literature is vast. For instance, [2] proposes a GRASP heuristic for the RSP that certainly needs to be referenced and discussed; while

© Springer International Publishing Switzerland 2016
M.J. Blesa et al. (Eds.): HM 2016, LNCS 9668, pp. 104–117, 2016.
DOI: 10.1007/978-3-319-39636-1_8

[1] contains a more recent evolutionary algorithm to address the RSP. We recommend the reader to consult [11] for further analysis and heuristics. Here, we propose a further generalization, summing-up all the previous ideas together. We study the Two-Node-Connected Star Problem (2NCSP), where the ring of the RSP is replaced by an arbitrary 2-node-connected component. This paper is organized as follows. Section 2 formally presents the 2NCSP and its computational complexity. Section 3 develops a GRASP heuristic enriched with a Variable Neighborhood descent and a Shaking operation. It is worth to mention that during some local searches we solve an integer linear programming model to find the best replacements in a feasible solution. Experimental results are provided in Sect. 4, while concluding remarks are presented in Sect. 5.

2 Two-Node-Connected Star Problem

We are given an undirected graph $G = (V, E)$ core-costs $\{c_e\}_{e \in E}$ and $\{d_e\}_{e \in E}$ for pendant links. The decision variable is a spanning subgraph $G' = (V, E')$, where $V = T \cup S$, being S the set of pendant nodes and T the terminal nodes from the 2-node-connected component. Each node $s \in S$ is pendant, so $deg_{G'}(s) = 1$, and the subgraph $H = (T, E'(T))$ is 2-node-connected. The goal in the 2NCSP is to find the subgraph G' at the minimum cost in the links meeting the previous constraints, where the cost-function is $c(G') = \sum_{e \in E'(T)} c_e + \sum_{e \in E - E'(T)} d_e$. Consider for example the feasible solution for the 2NCSP shown in Fig. 3. If $c_e = 1$ for all links in the core, and $d_e = 1$ for pendant links, then the cost is $c(G') 13 \times 1 + 18 \times 2 = 49$.

2.1 Computational Complexity

The 2NCSP is related to previous $\mathcal{NP}$-Hard combinatorial problems. The recognition of a Hamiltonian Tour belongs to one of the first 21 problems in $\mathcal{NP}$-Complete class [4]. The Traveling Salesman Problem belongs to this class, since it subsumes the recognition of a Hamiltonian tour in a graph. Recall the following result for the Minimum-Weight Two-Connected Spanning Problem, or MW2CSP for short [6]:

Lemma 1. *The MW2CSP belongs to the class of $\mathcal{NP}$-Complete problems*

Proof. By reduction to Hamiltonian Tour. Here we provide a sketch of the proof. If we are given an undirected graph $G = (V, E)$, then assign unit costs to every link $e \in E$, and double cost to links that do not belong to E. The cost for the optimal solution for the MW2CSP is not greater than $|E|$ if and only if G presents a Hamiltonian tour.

Theorem 1. *The 2NCSP belongs to the $\mathcal{NP}$-Complete class.*

Proof. A reduction to the MW2CSP is retrieved when pendant costs are infinite.

3   GRASP

Greedy Randomized Adaptive Search Procedure (GRASP) is a multi-start or iterative process, where feasible solutions are produced in a first phase, and neighbor solutions are explored in a second phase [8]. The best overall solution is returned as the result. This powerful metaheuristic has been succesfully implemented to address several problems, such as Internet Telephony, Cellular Systems, Connectivity, Cooperative Systems [10] and Wide Area Network design [9], Graph Theory and Combinatorics [3], among many others. It trades-off diversification and greediness, using randomization. Here, we will develop a GRASP heuristic enriched with a Variable Neighborhood Descent (VND) in order to address the 2NCSP. We invite the reader to consult [8] for a general GRASP template and further details.

3.1   Construction Phase for the 2NCSP

A feasible subgraph is produced whenever possible in our construction phase. It consists of five steps. First, we will describe each step. Then, a pseudocode for the construction phase will be provided. The following concept will be used:

Definition 1 (H-Path). *Given a subgraph H of G, an elementary path $P = \{p_1, \ldots, p_n\}$ in G is an H-Path if its intersection with H is precisely the terminal nodes from P: $P \cap H = \{p_1, p_n\}$.*

1. Three nodes i, j and k are selected in such a way that the area of the resulting triangle is maximized.
2. Three minimum-cost node-disjoint paths $P_{i,j}$, $P_{i,k}$ and $P_{j,k}$ are iteratively found. Figure 1 sketches an example, where nodes 6, 16 and 21 were randomly picked in Step 1, and the three paths $\{6, 0, 21\}$, $\{6, 16\}$ and $\{16, 24, 21\}$ are built in Step 2.
3. Nodes are iteratively added to the component H by means of H-Paths, whenever the resulting component does not have a prespecified number of nodes. The node v is picked uniformly at random from those nodes that do not

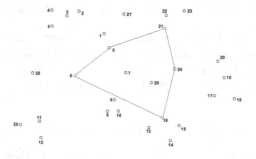

Fig. 1. First 2-node-connected component to build a feasible solution.

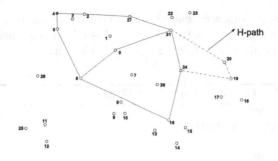

Fig. 2. Building an H-Path to connect node 19.

belong to H. Then, the shortest path between v and all the nodes from H are found, and the cheapest one is selected. A second disjoint shortest path is found, which, together with the first one, jointly determine an H-Path.

We illustrate this step in Fig. 2. Assume node 19 is selected. The shortest paths with all the nodes from the set $\{0, 6, 16, 21, 24\}$ are found. The cheapest one is path $\{19, 24\}$. Then, we consider the shortest paths with the nodes from the set $\{0, 6, 16, 21\}$. As a result, the path $\{19, 20, 21\}$ is selected. The resulting H-Path is $\{21, 20, 19, 24\}$.

4 Once we get a 2-node-connected component H with the prespecified number of nodes, the remaining nodes are connected (pendant to H) in a greedy fashion. Specifically, every node $v \notin H$ is connected to w such that $w = arg \min_{u \in H}\{d_{v,u} : (v, u) \in E\}$. If there is no potential link $(v, w) \in E$, there is no feasible solution such that H is the 2-node-connected component. In that case, we go to Step 1 again, in order to produce another solution. Figure 3 presents the pendant links in blue, and the links belonging to H in red. Observe that pendant nodes are assigned greedily to the closest nodes from H.

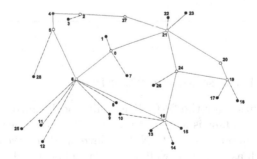

Fig. 3. After a greedy assignment, a feasible solution for the 2NCSP is produced.

5 Finally, irrelevant links are deleted, and a cost reduction is produced. A link e in a feasible solution G' is irrelevant if $G' - e$ is also a feasible solution. If we observe carefully Fig. 3, link $\{21, 24\}$ is irrelevant.

Algorithm 1. $G' = GreedyRandomized(G, Min)$

1: $H \leftarrow FirstCycle(G)$
2: $RCL \leftarrow Nodes(G) - Nodes(H)$
3: **while** $|H| < Min$ **do**
4:     $v \leftarrow Random(RCL)$
5:     $P_1 \leftarrow Dijkstra(v, G, H)$
6:     $P_2 \leftarrow Dijkstra(v, G, H, P_1)$
7:     $H \leftarrow H \cup P_1 \cup P_2$
8:     $RCL \leftarrow RCL - \{P_1, P_2\}$
9: **end while**
10: $G' \leftarrow H$
11: **while** $|RCL| > 0$ **do**
12:     $v \leftarrow Random(RCL)$
13:     $w \leftarrow arg \min_z \{d_{v,z}\}$
14:     $G' \leftarrow G' \cup \{v, w\}$
15:     $RCL \leftarrow RCL - \{v\}$
16: **end while**
17: **return** G'

Algorithm 2. $H = FirstCycle(G)$

1: **if** $RandomBoolean = 1$ **then**
2:     $\{i, j, k\} \leftarrow Random(3, G)$
3: **else**
4:     $\{i, j, k\} \leftarrow LargestArea(3, G)$
5: **end if**
6: $P_{i,j} \leftarrow Dijkstra(G, i, j)$
7: $P_{j,k} \leftarrow Dijkstra(G, j, k, P_{i,j})$
8: $P_{i,k} \leftarrow Dijkstra(G, i, k, P_{i,j} \; cupP_{j,k})$
9: $H \leftarrow P_{i,j} \cup P_{i,k} \cup P_{j,k}$
10: **return** H

3.2   Local Search Phase for the 2NCSP

Once a feasible solution for the 2NCSP is built, local improvements take place in a GRASP heuristic. This is a second phase, called Local Search. In this work, a Variable Neighborhood Descent (VND) is developed [7]. We consider classical neighborhoods (such as 2-opt), as well as other neighborhoods that exploit exact linear integer programming solutions.

The flow chart presented in Fig. 4 corresponds to the VND developed as the Local Search phase of our GRASP methodology. We define six neighborhoods,

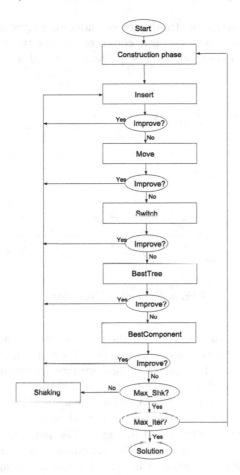

Fig. 4. Variable Neighborhood Descent (VND). Flow diagram.

which will be fully detailed in the following paragraphs. It is worth to remark that a Shaking operation is produced at the end, where the solution is perturbed in order to run out from local minima once the process does not accept a better local solution. We remark that all local searches preserve feasibility.

Local Search 1: Insert. The goal of this local search is to reconnect a pendant node and include it into the 2-node-connected component in the best manner.

Definition 2. *Given an instance of 2NCSP and a feasible solution G', the neighborhood of G' is the set of feasible solutions G'' such that inserts a pendant node from G' inside its 2-node-connected component.*

Once an insertion to the 2-node-connected component takes place, all the pendant nodes should be greedily linked again, since the best assignment could have been modified.

Two different insertion methods are considered. Figure 5 presents the first method, where the pendant node u is connected to the two closest nodes v and w from the connected component. The paths $u - v$ and $u - w$ are found using Dijkstra algorithm.

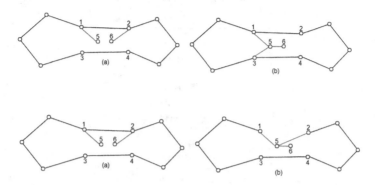

Fig. 5. Example of insertion.

In our example, we choose $u = 5$ as the pendant node, and the closest nodes are $v = 1$ and $w = 3$. Direct links (u, v) and (u, w) are drawn (even though a path is a general connection). This method may produce irrelevant links. Therefore, a deletion of possible irrelevant links takes place. If there is no paths between u and some of the nodes v and w, the insertion is not feasible, u is not a candidate for insertion, and another pendant node is considered.

The second method is illustrated in Fig. 5. The pendant node u is inserted between two incident nodes from the 2-node-connected component. As in the first method, the remaining pendant nodes are reconnected again. No irrelevant links are produced.

In the example we pick $u = 5$ as the pendant node, and $(v, w) = (1, 2)$ is the link from the 2-node-connected component. Links $(u, v) = (5, 1)$ and $(u, w) = (5, 2)$ are included, and $(v, w) = (1, 2)$ is deleted. Finally, the other pendant nodes are reconnected again (node 6 is reconnected to node 5). This insertion test is iteratively applied to each pendant node, and finally the best insertion is introduced.

Local Search 2: Move. The idea is to extract a specific node from the 2-node-connected component and move it in another position of this component, in a feasible way.

Definition 3. *Given an instance for the 2NCSP and a feasible solution G', we define as a neighbor any feasible solution that moves an arbitrary node from one position to another, in a feasible way.*

The target node u from the 2-node-connected component has degree two. In a first step, u is removed and its incident nodes are connected (if there is node link between the incident nodes, another node is selected instead). In a second step, u is re-inserted analogously. Figure 6 illustrate Step 1, where the node $u = 5$ is removed and the adjacent nodes 1 and 2 are reconnected. Figure 7 show Step 2, where u is reinserted to nodes 3 and 4.

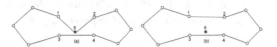

Fig. 6. Example of move. Step 1.

Fig. 7. Example of move. Step 2.

After all feasible reinsertions are studied, the cheapest one is selected. A new node from the component is selected until all nodes are tested. Finally, the best movement is applied to the solution.

Local Search 3: Switch. The goal is to switch a node from the main component into a pendant one, in the best way.

Definition 4. *Given an instance for the 2NCSP and a feasible solution G', we define as a neighbor any feasible solution that extracts a node from the 2-node-connected component and assign it greedily as a pendant node.*

The extraction method is identical to the Local Search 2 (Move). Recall that the extracted node must have degree 2, and their adjacent nodes must share a link. Then, the node is greedily assigned as pendant to the 2-node-connected component. The process is iteratively performed in each feasible extraction. Only the extraction with the highest cost reduction is produced.

Local Search 4: BestTree. The goal of this local search is to replace an elementary path with pendant nodes by the best path composed by the same nodes. The best path is found by means of an integer linear programming model.

Definition 5 (Path with pendant nodes). *Let $G = (V, E)$ an undirected graph. We say $C = (V', E')$ is a path with pendant nodes and extremes u and v if and only if the following conditions are met:*

– $p(u, v)$ *is an elementary path that connects the nodes* u *and* v *in* G.
– C *is an induced subgraph of* G. *It has all nodes from* $p(u, v)$ *and the pendant nodes of them.*
– C *is a tree.*

Definition 6. *Given an instance for the 2NCSP and a feasible solution* G', *we define as a neighbor any feasible solution that replaces a path with pendant nodes by another.*

The idea is to find a path with pendant nodes and replace it by the best path using an integer linear programming model.

Figure 8 presents a pictorial example. Suppose that the selected path is $\{6, 5, 4, 2, 24, 21\}$ (see Fig. 9). We count the number of nodes, and add the number of pendant nodes. If it exceeds a certain threshold, the path is discarded. The reasons to discard the path is that an exact solution for the integer linear programming is computationally prohibitive for large instances. Figure 10 presents the best path using those nodes. Once we get the best path with pendant nodes, we replace the path by the best one, whenever the cost is lower. Figure 11 shows the new feasible solution for the 2NCSP.

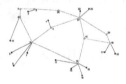

Fig. 8. Input network.

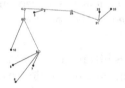

Fig. 9. Selected path

Fig. 10. Best tree

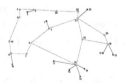

Fig. 11. Replacement

Local Search 5: BestComponent. The idea is to replace cycles by the optimum 2-node-connected component using its nodes. It is worth to remark that the resulting topology is not necessarily a cycle.

Definition 7. *Given an instance for the 2NCSP and a feasible solution G', we define as a neighbor any feasible solution that is a replacement of a cycle by a 2-node-connected graph with the same nodes and lower cost.*

This local search applies three operations that lead to a cost reduction in the cycles of a feasible solution. They are based on integer linear programming and swaps.

Swap. Swap or 2-opt was proposed in 1958 by Croes in order to solve the Traveling Salesman Problem (TSP). The idea is to modify the order in which the Hamiltonian tour is built, in order to avoid crosses. In our particular problem, 2-opt receives a cycle and returns the best swap between two nodes. The output is another cycle where nodes i and j are visited in different in a reverse order.

Shaking. The goal of this operation is to run out from local optima, by means of random perturbations to the current solution. A local optima is met when there is no feasible improvement under all neighborhood structures. In this case, *Shaking* is applied to the solution, and the Local Search phase takes place again. It picks a random number of nodes from the 2-node-connected component, deletes them and assigns them as pendant nodes. The resulting component should have at least 3 nodes. If it is not possible to assign the deleted nodes, the shaking process is repeated.

Algorithm 3. $G' = Shaking(G)$

1: $H \leftarrow 2Component(G)$
2: $Discard \leftarrow Random(0, |H| - 3)$
3: **for** $i = 1 : Discard$ **do**
4:     $v \leftarrow Random(H)$
5:     $H \leftarrow H - \{v\}$
6: **end for**
7: $G' \leftarrow AddpendantNodes(G, H)$
8: **return** G'

---

4   Experimental Results

As far as we know, there is no literature in the field related with 2NCSP. As a consequence, there is no source in order to contrast the results of our algorithm. However, there are relaxations of our problem, for instance, the Ring Star Problem (RSP), where rings are used instead of general 2-node-connected components as a core. We consider M. Labbé et al. as a reference study for the RSP [5]. Our tests were developed in an Intel i7 processor with 4GB Ram. All the algorithms, data structures and graph operations were developed with no libraries. The exact resolution of local searches were validated using AMPL modelling language and using CPLEX 12.5 solver. IBM Concert Technology was considered for the integer linear programming model in C++.

4.1 Test Cases

M. Labbé et al. define three classes of test cases in order to evaluate their algorithm performance for the RSP solver. Class I consists of TSPLIB 2.1 instances, ranging from 50 to 200 nodes. This class will be considered to evaluate our heuristic. Now, let us describe the way the authors in [5] process the test cases. Let l_{ij} be the Euclidean distance between the nodes v_i and v_j. The connection cost is $c_{ij} = \alpha \times l_{ij}$, while the pendant cost is $d_{ij} = (10 - \alpha) \times l_{ij}$, being $\alpha \in \{3, 5, 7, 9\}$. The authors provide a set of 27 graphs from TSPLIB and different values $\alpha \in \{3, 5, 7, 9\}$, defining 108 instances. Our GRASP/VND heuristic was executed 100 times per instance using 20 iterations in the Shaking process.

4.2 Results

Curiously enough, our GRASP/VND heuristic produces Ring Star topologies in all instances. As a consequence, the comparison is fair, since they are all feasible solutions for the RSP. Table 1 presents the instances where our GRASP/VND solution outperforms the reference study [5].

- Instance: the code of the TSPLIB 2.1 instance.
- α: the weighting factor of connection and assignment costs.
- opt_L: cost found from Labbé et al. reported in [5].
- obt_G: cost found by our GRASP heuristic.
- $gap = (opt_G - opt_L)/opt_L \times 100$.
- t: processing time (expressed in sexagesimal mm:ss).

Our algorithm outperforms the one from Labbé et al. in 43 out of 108 instances. The same solution is produced in 50 instances, while the best results so far for the remaining 15 instances are produced by Labbé et al. algorithm. The CPU time of our GRASP resolution is greater than two minutes only for instances with hundreds of nodes. On the other hand, the report from Labbé shows a CPU time of more than two hours in some instances. Table 2 presents the average gaps among all the instances considered in [5].

It is worth to remark the GRASP/VND performance when $\alpha = 9$, this is, when pendant links are cheaper than connection links. The number of pendant nodes is monotonically increasing with respect to this weighting factor α.

The integer linear programming models are not suitable for subgraphs with more than 15 nodes, since the computational effort is prohibitive. Further studies lead to the conclusion that the quality of the solution is highly sensitive to the Local Search phase, and the order of the searches rather than the construction phase. The order of the VND has been tuned in a preliminary stage. Additionally, the local searches based on integer linear programming solutions play a key role in the quality of the solutions, which are in fact competitive with the authoritative work from Labbé et al. Finally, the shaking operation is meaningful in the results, since it has been called several times before the best solution is returned. In spite of the diversification introduced by different local searches, we could check that the second phase retrieves a local optima for several instances.

Table 1. Comparison between Labbé et al. algorithm and GRASP/VND.

Instance	α	opt_L	opt_G	gap	t
eil51	9	1.244	1.224	-1,61	0:07
berlin52	7	37.376	37.029	-0,93	0:10
berlin52	9	20.361	19.887	-2,33	0:05
brazil58	9	83,690	81,824	-2,23	0:04
st70	9	2.610	2.513	-3,72	0:32
eil76	9	1.710	1.690	-1.17	0:37
pr76	9	424.359	378.396	-10,83	0:31
gr96	9	232,823	231,954	-0,37	1:15
rat99	7	6.436	6.301	-2,05	1:30
rat99	9	5.150	4.655	-9.61	1:25
kroc100	7	113.533	112.764	-0,68	1:30
kroc100	9	92.894	91.451	-1.55	1:25
krod100	7	116.849	116.715	-0.11	1:28
krod100	9	92.102	90.615	-1.61	1:25
rd100	7	40.915	40.912	-0,01	1:40
rd100	9	31.776	31.317	-1,44	1:19
lin105	5	69.365	69.354	-0,02	> 2:00
lin105	9	96.920	63.277	-9,50	> 2:00
pr107	5	210.465	210.153	-0,15	> 2:00
pr107	7	259.571	258.191	-0,53	> 2:00
pr107	9	264.918	258.389	-2,46	> 2:00
gr120	9	24,322	23,503	-3,36	> 2:00
pr124	9	340.153	323.764	-4.82	> 2:00
bier127	9	347.845	343.893	-1,14	> 2:00
ch130	5	28.790	28679	-0,39	> 2:00
ch130	7	32.707	32.499	-0,64	> 2:00
ch130	9	23.639	23.657	-2.37	> 2:00
pr136	7	491.981	491.296	-0,14	> 2:00
pr136	9	387.327	378.136	-2,73	> 2:00
gr137	9	335,009	333,49	-0,45	> 2:00
pr144	7	383.041	382.827	-0,06	> 2:00
pr144	9	366.833	357.958	-2.42	> 2:00
ch150	9	26.371	25.777	-2,25	> 2:00
kroa150	9	113.080	111.882	-1,06	> 2:00
krob150	9	108.885	107.925	-0,88	> 2:00
pr152	5	376.155	364.406	-3,12	> 2:00
pr152	7	475.052	463.221	-2,49	> 2:00
pr152	9	475.440	461.264	-2,98	>2:00
rat195	9	9.395	8.585	-8.62	> 2:00
d198	9	97,899	97,224	-0,69	> 2:00
kroa200	3	93.699	90.999	-2,88	> 2:00
kroa200	9	124.678	122.809	-1,50	> 2:00
krob200	9	127.800	124.086	-2,91	> 2:00

Table 2. Performance of our GRASP/VND heuristic.

α	$\overline{gap}$
3	0.57
5	0.29
7	-0.02
9	-2.65

5   Conclusions

A network optimization problem, called 2-Node-Connected Star Problem (2NCSP), is here studied. It is a relaxation of the Ring Star Problem (RSP), where the core can be an arbitrary 2-node-connected component. This selection is inspired in the 4/3-factor proposed in a foundational article in structural network design, authored by Clyde Monma et al. The 2NCSP is precisely the minimum weighted two connectivity when pendant costs are infinite. Therefore, it belongs to the class of $\mathcal{NP}$-Hard computational problems. No polynomial time exists, unless $\mathcal{P} = \mathcal{NP}$. For that reason, a GRASP/VND metaheuristic is here developed. Test cases were considered using TSPLIB adapted by Labbé et al. Even though the core is optimized during the heuristic, the final result was a ring core under all instances. As a consequence, the comparison using those RSP instances is fair. The proposed heuristic outperforms previous solutions, mainly when pendant nodes are cheaper than nodes connected in the core. A concluding remark is that our relaxation could not take effect in practice, even though it is supported theoretically. However, the best results so far in particular RSP instances were obtained as a bonus. The replacements of the best core and best paths make the difference, even though it produces rings. We believe the corresponding neighbourhood structures are the main ingredient of the proposed heuristic.

As future work, we would like to deploy real-life solutions using the 2NCSP as a reference model.

References

1. Calvete, H.I., Gal, C., Iranzo, J.A.: An efficient evolutionary algorithm for the ring star problem. Eur. J. Oper. Res. **231**(1), 22–33 (2013)
2. Dias, T.C.S., de Sousa Filho, G.F., Macambira, E.M., dos Anjos F. Cabral, L., Fampa, M.H.C.: An efficient heuristic for the ring star problem. In: Àlvarez, C., Serna, M. (eds.) WEA 2006. LNCS, vol. 4007, pp. 24–35. Springer, Heidelberg (2006)
3. Festa, P., Pardalos, P., Resende, M., Ribeiro, C.: Randomized heuristics for the max-cut problem. Optim. Methods Softw. **17**(6), 1033–1058 (2002)
4. Karp, R.M.: Reducibility among combinatorial problems. In: Miller, R.E., Thatcher, J.W. (eds.) Complexity of Computer Computations, pp. 85–103. Plenum Press (1972)

5. Labbé, M., Laporte, G., Martín, I.R., González, J.J.S.: The ring star problem: polyhedral analysis and exact algorithm. Networks **43**(3), 177–189 (2004)
6. Monma, C.L., Munson, B.S., Pulleyblank, W.R.: Minimum-weight two-connected spanning networks. Math. Program. **46**(1–3), 153–171 (1990)
7. Pérez, J.A.M., Moreno-Vega, J.M., Martín, I.R.: Variable neighborhood tabu search and its application to the median cycle problem. Eur. J. Oper. Res. **151**(2), 365–378 (2003)
8. Resende, M., Ribeiro, C.: GRASP: greedy randomized adaptive search procedures. In: Burke, E.K., Kendall, G. (eds.) Search Methodologies, pp. 287–312. Springer, US (2014)
9. Robledo Amoza, F.: GRASP heuristics for Wide Area Network design. Ph.D. thesis, INRIA/IRISA, Université de Rennes I, Rennes, France (2005)
10. Romero, P.: Mathematical analysis of scheduling policies in Peer-to-Peer video streaming networks. Ph.D. thesis, Facultad de Ingeniería, Universidad de la República, Montevideo, Uruguay, November 2012
11. Simonetti, L., Frota, Y., de Souza, C.: The ring-star problem: a new integer programming formulation and a branch-and-cut algorithm. Discrete Appl. Math. **159**(16), 1901–1914 (2011). 8th Cologne/Twente Workshop on Graphs and Combinatorial Optimization (CTW 2009)

Neighborhood Composition Strategies in Stochastic Local Search

Janniele A.S. Araujo[1,2(✉)], Haroldo G. Santos[2], Davi D. Baltar[1],
Túlio A.M. Toffolo[2,3], and Tony Wauters[3]

[1] Computer and Systems Department,
Federal University of Ouro Preto, Ouro Preto, Brazil
janniele@decsi.ufop.br, davibaltarx@gmail.com
[2] Department of Computing, Federal University of Ouro Preto, Ouro Preto, Brazil
haroldo@iceb.ufop.br, tulio@toffolo.com.br
[3] Computer Science Department, CODeS, KU Leuven, Leuven, Belgium
tony.wauters@cs.kuleuven.be

Abstract. Methods based on Stochastic Local Search (SLS) have been ranked as the best heuristics available for many hard combinatorial optimization problems. The design of SLS methods which use many neighborhoods poses difficult questions regarding the exploration of these neighborhoods: how much computational effort should be invested in each neighborhood? Should this effort remain fixed during the entire search or should it be dynamically updated as the search progresses? Additionally, is it possible to learn the best configurations during *runtime* without sacrificing too much the computational efficiency of the search method? In this paper we explore different tuning strategies to configure a state-of-the-art algorithm employing fourteen neighborhoods for the Multi-Mode Resource Constrained Multi-Project Scheduling Problem. An extensive set of computational experiments provide interesting insights for neighborhood selection and improved upper bounds for many hard instances from the literature.

Keywords: Stochastic local search · Project scheduling · Multi-neighborhood search · Learning · Online and offline tuning

1 Introduction

Stochastic Local Search (SLS) methods obtained best results for many optimization problems. In school timetabling, Fonseca et al. [3] won the Third International Timetabling competition with a hybrid Simulated Annealing incorporating eight neighborhoods. In Project Scheduling, Asta et al. [1] won the MISTA Challenge [9] with a Monte Carlo based search method with nine neighborhoods. In both methods, neighborhoods are explored stochastically instead of the much popular deterministic best/first fit alternatives where local optimality is usually reached at every iteration. Instead, in these SLS algorithms, a random neighbor is generated in one of the neighborhoods and its acceptance is immediately

© Springer International Publishing Switzerland 2016
M.J. Blesa et al. (Eds.): HM 2016, LNCS 9668, pp. 118–130, 2016.
DOI: 10.1007/978-3-319-39636-1_9

decided. In this paper, we focus on SLS algorithms with these characteristics, instead of considering the broader definition of SLS [6].

Reasons for the good performance of these methods are subject of study. Some insights may come from theoretical models such as [7] or from experimental studies such as [5]. As both studies point out, more connected search spaces provide a landscape where it is easier to escape from attractive basins in the search landscape. This is a very favorable point of SLS, since the different neighborhoods and the flexibility used in their exploration contribute to an increased connectivity of the search space graph.

Given a set of neighborhoods $\{\mathcal{N}_1, \ldots, \mathcal{N}_k\}$ and a search method, we define probabilities $\mathbf{p} = (p_1, \ldots, p_k)$ of selecting each neighborhood at each iteration as *neighborhood composition*. As k increases, it gets harder to find the best configuration of $\mathbf{p}$. Some important considerations are: should $\mathbf{p}$ remain fixed during the entire search or should it be dynamically updated as the search progresses? Is it possible to learn the best configurations during runtime without sacrificing too much the computational efficiency of the search method? In this paper we try to answer these questions considering a state-of-the-art heuristic for the Multi-Mode Resource Constrained Multi-Project Scheduling Problem where fourteen neighborhoods are stochastically explored.

2 The Multi-Mode Resource Constrained Multi-Project Scheduling Problem

The Multi-Mode Resource Constrained Multi-Project Scheduling Problem (MMRCMPSP) is a generalization of the Project Scheduling Problem (PSP). In this problem jobs can be processed in different modes, with varying execution speeds and consuming different amounts of resources. Both renewable and non-renewable resources are present. Note that the non-renewable resources render the generation of an initial feasible solution NP-Hard, since selecting a valid set of modes corresponds to solving a multi-dimensional knapsack problem. The objective function considers the project delays.

An instance of the MMRCMPSP [9] can be defined by:

P projects with release dates h_p, $p \in P$;
J jobs, connected by a set of precedence relations B;
K nonrenewable resources with capacities o_k, $k \in K$;
R renewable resources with capacities q_r, $r \in R$;

Jobs must be scheduled respecting release dates and precedence constraints. Each job has to be assigned to a processing mode, which defines its resource consumption and duration. A job j executed in mode m has duration d_{jm} and consumes u_{kjm} units of non-renewable resource k, $k \in K$, and v_{rjm} units of renewable resource r, $r \in R$. The objective functions minimize, hierarchically: (*i*) the total project delay (TPD), which considers project completion times and (*ii*) the total makespan (TMS), which considers the completion of the last project.

3   Neighborhoods

A solution is represented by an ordered pair $(\pi, \mathcal{M})$, where π indicates an allocation sequence and $\mathcal{M}$ is a feasible set of modes, i.e. an allocation of modes which respects the availability of nonrenewable resources. A complete solution is decoded with a Serial SGS algorithm [4] from $(\pi, \mathcal{M})$, allocating each job in the first timeslot where precedence constraints are respected and sufficient renewable resources are available. All neighborhoods operate in the search space of feasible values for $(\pi, \mathcal{M})$, i.e. movements violating precedence constraints or nonrenewable resource usage constraints are discarded. Neighborhoods operate either on π or $\mathcal{M}$. Most neighborhoods were proposed by [1]. Some of these neighborhoods were conceived aiming at the minimization of the TPD. They *compact* jobs in π, reorganizing them so that all jobs of a given project are contiguous in the allocation sequence.

Fourteen neighborhoods are considered:

SPE squeeze project on extreme: all jobs of a project p are compacted around a reference job, while jobs of other projects are moved either before or after project p; Fig. 1 shows a sample neighbor s' of s in the SPE neighborhood, with position 4 of π being used as a reference job, thus all jobs pertaining to the same project of job 14 are squeezed immediately at its side;

SCP swap and compact two projects: relative positions of projects are swapped and all jobs of both projects are sequentially allocated, starting with jobs from the project which was previously starting at a latter timeslot;

OP offset project: all jobs of a given project are shifted by the same number of positions;

CPP compact project on percentage: a percentage of jobs of a project p are justified to the left, starting from the end of π; Fig. 2 shows a sample neighbor s' of s in the CPP neighborhood considering project 4 (jobs shown in shaded cells) and compaction percentage 0.5; note that the project has 5 jobs, so $\lfloor 5 \times 0.5 \rfloor = 2$ and thus the last two jobs are contiguously allocated in π;

ISJ invert sequence of jobs: a subsequence of jobs in π is inverted;

OJ offset job: shifts the position of one job in the sequence π;

STJ swap two jobs: swaps the position of two jobs in the sequence;

CSP compact subsequent projects: compress a contiguous list of projects in a sequence;

SSJW successive swap of a job in a window: all possible swap positions for a job in a window are explored in a first-fit fashion;

$$s'=\text{SPE}(s,4)$$

s	2	3	5	8	14	10	7	9	12
s'	3	8	2	5	14	10	9	7	12

Fig. 1. Example of a neighbor s' generated in the SPE neighborhood of s

SIJW **successive insertions of a job in a window:** similar to SSJW, but instead
of swapping jobs, a job is re-inserted in another position within the explored
window;

C1M **change one mode:** changes the mode of one job;

C2M **change two modes:** changes the mode of two jobs;

C3M **change three modes:** changes the mode of three jobs; to reduce the size
of this neighborhood only triples of consecutive jobs in the precedence graph
are considered;

C4M **change four modes:** changes the mode of four jobs that are consecutive
in the precedence graph (similarly to C3M).

$$s'=CPP(s,4,0.5)$$

s	2	3	5	8	14	10	7	9	12
s'	2	3	5	8	14	10	9	7	12

Fig. 2. Example of a neighbor s' generated in the CPP neighborhood of s

4   Offline Neighborhood Composition

In this section we evaluate a metric to define **p** *a priori*, i.e. based on a statistical
analysis of the neighborhoods efficiency. The performance of all neighborhoods
in a previously generated pool of solutions is considered. We call a neighborhood *efficient* when its stochastic exploration produces good results, i.e. if it
improves the solution cost or generates sideways moves[1] with minimal computational effort.

In a preliminary set of experiments we observed that the efficiency of different neighborhoods varied considerably depending on the *stage* of the search.
Neighborhoods which were quite efficient in the beginning of the search did not
present the same good properties as the search advanced.

We evaluate the efficiency of neighborhoods in two different stages. In the
first stage, low quality (initial) solutions are considered, generated by quick constructive algorithms. In the second stage, incumbent solutions obtained after
10,000 iterations without improvement of a state-of-the-art LAHC solver [8] are
selected. Table 1 describes the minimum, maximum and average solution quality[2]
for solutions of both phases. Each value consider all instances and 10 solutions
per instance (300 solutions in total), taking into account the best known solutions in literature to compute the quality.

The efficiency of each neighborhood k was computed considering a random
sampling of neighbors (10,000) for each solution in the pool. It was observed that

---

[1] Moves which do not change the objective function value but modify the solution.

[2] The quality $q_i = \frac{b_i}{c_i}$ of a solution for instance i with cost c_i, considering the best
known solution's cost b_i.

Table 1. Characteristics of the solution pool used for *offline* neighborhood analysis

	Solution quality		
	Minimum	Average	Maximum
0	<0.0057	0.3566	0.6666
10000	0.6175	0.8193	1.000

only analyzing the improvement of neighbors was not enough, so we consider the number of neighbors corresponding to improved solutions $\tilde{s}(k)$, the number of neighbors corresponding to sideways moves $\underline{s}(k)$ with factor $\vartheta \in [0,1]$, and the total CPU time $\tilde{c}(k)$ spent validating and evaluating neighbors. Thus, the efficiency $\tilde{e}_k$ of a neighborhood k is given by:

$$\tilde{e}_k = \begin{cases} \dfrac{\tilde{s}(k) + \vartheta\,\underline{s}(k)}{\tilde{c}(k)} & \text{if } \tilde{c}(k) > 0 \\ 0 & \text{otherwise} \end{cases} \tag{1}$$

To ease comparison, we normalize the efficiency values. Let $\max(\tilde{e})$ be the maximum $\tilde{e}_k$ for all $k \in N$. Equation (2) shows how the normalized efficiency e_k is obtained for a neighborhood $k \in N$:

$$e_k = \frac{\tilde{e}_k}{\max(\tilde{e})} \tag{2}$$

Table 2. Normalized neighborhoods efficiency computed *offline*, considering the two stages solution pool

$\vartheta = 0.0$				$\vartheta = 0.1$			
First stage		Second stage		First stage		Second stage	
C1M	1.0000	C1M	1.0000	C1M	1.0000	OJ	1.0000
OP	0.7407	C2M	0.2990	OJ	0.6969	CSP	0.7323
C2M	0.7288	OJ	0.2609	OP	0.6661	STJ	0.6651
OJ	0.5143	STJ	0.2076	C2M	0.6427	C1M	0.5709
CPP	0.4729	ISJ	0.2037	STJ	0.4650	ISJ	0.4345
C3M	0.4716	OP	0.1812	C3M	0.4091	OP	0.3474
STJ	0.3677	C3M	0.0950	CPP	0.4053	CPP	0.1959
SCTP	0.3281	SPE	0.0540	ISJ	0.2824	SCTP	0.1921
C4M	0.3058	SCTP	0.0502	SCTP	0.2753	SPE	0.1789
SPE	0.2725	C4M	0.0167	C4M	0.2578	C2M	0.0865
ISJ	0.2304	CPP	0.0023	CSP	0.2540	C3M	0.0371
SSJW	0.0024	SSJW	0.0019	SPE	0.2327	C4M	0.0029
SIJW	0.0005	SSIW	0.0003	SSJW	0.0016	SSJW	0.0001
CSP	0.0000	CSP	0.0000	SIJW	0.0000	SIJW	0.0000

Table 2 presents an efficiency comparison between all neighborhoods in the two different stages. To illustrate the impact of the sideways moves on the final solution quality, the first columns are computed with $\vartheta = 0$ and the last ones are computed with $\vartheta = 0.1$. Note that neighborhoods which operate on entire projects, like OP, CPP, SCTP, are usually well ranked in the first stage. Their efficiency dramatically decreases in the second stage, except for the CSP, that has a very low efficiency in the first stage, increasing in the second one due mainly to sideways moves. Neighborhoods which change one mode, swap, inverts or shifts jobs are the most significant for the second stage, corresponding to a stage of small adjustments in π and $\mathcal{M}$.

5 Online Neighborhood Composition

In this section we consider the *online* neighborhood composition. In this approach, all neighborhoods start with the same probability of being chosen and this probability is dynamically updated considering results obtained during the search. While this approach introduces a learning overhead, it can more

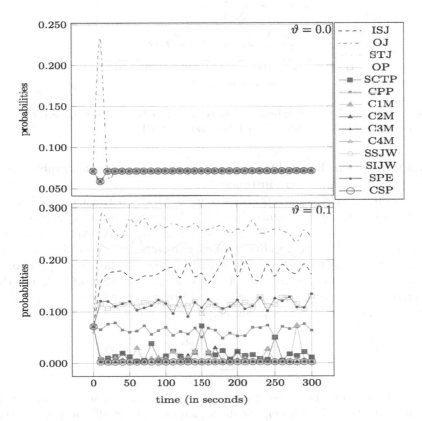

Fig. 3. Evolution of the neighborhood selection probabilities over time considering instances A-10 and *online* tuning ($z = 1,000$ and $\beta = 0.01$)

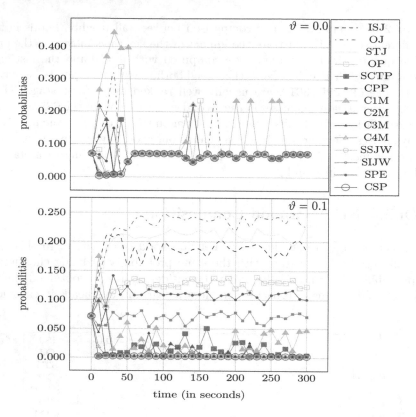

Fig. 4. Evolution of the neighborhood selection probabilities over time considering instances B-9 and *online* tuning ($z = 1,000$ and $\beta = 0.01$)

easily adapt itself if the neighborhoods' efficiencies change considerably during the search or vary in different instances.

Whenever a sample of z neighbors is explored, the normalized efficiency e_k (Eqs. (1) and (2)) of each neighborhood $k \in N$ is computed and their selection probabilities are updated according to the results obtained exploring this last batch of neighbors. The probability p_k of selecting a neighborhood $k \in N$ is given by Eq. (3). Note that a constant β is included when calculating the probability. This constant prevents assigning probability zero to a neighborhood, and thus all neighborhoods have a minimal chance of being selected in any stage of the search. All experiments in this paper employs $\beta = 0.01$.

$$p_k = \frac{e_k + \beta}{\displaystyle\sum_{k' \in N} (e_{k'} + \beta)} \tag{3}$$

The definition of z is a crucial point: while larger values of z provide a more accurate evaluation of each neighborhood, smaller values offer a more reactive method.

Figure 3 shows how the probabilities evolve over time considering a small instance, A-10. The first graph assumes $\vartheta = 0.0$ (sideways moves are not rewarded), while the second considers $\vartheta = 0.1$ (sideways moves are partially rewarded). Note that, as one would expect, most improving moves for small instances are executed in the beginning of the search. Therefore, once a local optima is reached, the probabilities remain unchanged when sideways moves are not considered. When sideways moves are considered, however, the probabilities vary during the entire search. It is noteworthy that more complex and expensive neighborhoods such as CSP, SSJW and SIJW were given lower probabilities when $\vartheta = 0.1$.

Figure 4 presents similar graphs to those of Fig. 3, but considering a larger instance, B-9. For this instance, neighborhoods C1M and C2M, which change job execution modes, were highly rewarded when $\vartheta = 0.0$. This stresses the importance of the modes selection for this particular instance. Additionally, the simple neighborhoods OJ and OP, which shifts the position of one job and one project, respectively, were also assigned high probabilities. Note that if $\vartheta = 0.0$ and the current solution is not improved by any neighborhood within z iterations,

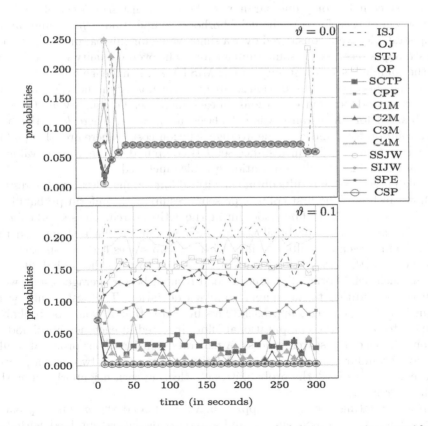

Fig. 5. Evolution of the neighborhood selection probabilities over time considering instances X-10 and *online* tuning ($z = 1,000$ and $\beta = 0.01$)

all neighborhoods are assigned the same probability. Figure 4 shows that this situation is recurring.

When sideways moves are rewarded ($\vartheta = 0.1$), the neighborhoods ISJ, OJ and STJ, which change the sequence of jobs, were given high selection probabilities during the entire search.

Figure 5 presents how probabilities evolve for a medium size instance, X-10. We can see that neighborhoods C1M, OJ, OP remain the most significant ones. Other neighborhoods appear with high probabilities at the beginning, like C2M. When $\vartheta = 0.1$ the neighborhoods with higher potential to generate sideways moves are assigned higher probabilities.

6   Comparing Composition Approaches

We evaluated the performance of *offline* and *online* tuning strategies in a meta-heuristic framework consisting of the Late Acceptance Hill Climbing (LAHC) algorithm. At each iteration a random neighborhood is chosen considering the probabilities **p** in a roulette scheme [2]. The LAHC metaheuristic suits well our purposes since it has only one parameter, the LAHC list size. Most of our tuning effort could then focus in the neighborhood composition. *Offline* and *online* tuning strategies were considered over a single value for this parameter[3], so that both strategies receives the same tuning effort. The overall quality was evaluated over the complete set of instances of the MISTA 2013 challenge [9].

Each neighborhood composition strategy was evaluated on the whole instance set using five independent executions on each instance. The quality q_i of a solution obtained for an instance i is calculated as $q_i = \frac{b_i}{a_i}$, where b_i is the best known solution for i and a_i is the average solution costs. Therefore, the quality $\omega \in [0, 1]$ of solutions for an instance set I is computed as the average value of q_i for all $i \in I$. The standard deviation σ is also included.

Table 3 presents the results obtained with *offline* tuning. Best average results are shown in bold. The first column presents results with uniform probabilities ($1/k$) for selecting all neighborhood during the entire search. The second column shows the results when probabilities are defined using a single stage, and the third column presents results obtained with the two stages tuning approach. As it can be seen, the efficiency metric (Eq. (1)) used to tune the probabilities of selecting each neighborhood produced good results: better average results were produced, in addition to an smaller standard deviation. The best results were obtained in the two stages approach. This indicates that it would be probably beneficial to update the computational effort invested in each neighborhood in different phases of the search process. A natural extension of our proposal would be a more granular approach, i.e. the definition of more than two search phases to more properly adjust the probabilities of selecting each neighborhood as the search progresses.

After evaluating the *offline* approach, we evaluated the *online* approach. Table 4 shows the average quality and standard deviation obtained with the

---

[3] LAHC list size value was fixed to 1,000.

Table 3. Quality of results for the *offline* tuning strategy

Offline tuning	Uniform	Single stage	Two stages
$\omega_{\vartheta=0}$	0.935	0.949	**0.955**
$\sigma_{\vartheta=0}$	0.041	0.039	0.039
$\omega_{\vartheta=0.1}$	0.935	0.952	0.953
$\sigma_{\vartheta=0.1}$	0.041	0.038	0.037

Table 4. Quality of results for *online* tuning $\beta = 0.01$

Online tuning	$z{=}1000$	$z{=}10000$	$z{=}50000$	$z{=}100000$
$\omega_{\vartheta=0}$	**0.9468**	0.9428	0.9431	0.9462
$\sigma_{\vartheta=0}$	0.0438	0.0415	0.0444	0.0437
$\omega_{\vartheta=0.1}$	0.9339	0.9358	0.9309	0.9330
$\sigma_{\vartheta=0.1}$	0.0420	0.0422	0.0411	0.0428

online approach using the metric $\tilde{e}$ and different values of z (interval in which probabilities are updated). The best results were obtained with $z = 1000$ and $\vartheta - 0$.

7   Composition Strategies Results

This section presents the results (Table 5) obtained by the LAHC metaheuristic considering both neighborhood composition strategies, and compared them with the best results from literature. An Intel ® Core i7-4790 processor with 3.6 GHz and 16 GD of RAM running SUSE Leap Linux was used during the experiments. All algorithms were coded in ANSI C 99 and compiled with GCC 4.8.3 using flags *-Ofast* and *-flto*. The runtime limit was set to 300 s (the same used in the MISTA Challenge [9]).

Table 5 shows the results obtained with the LAHC implementation and the best results obtained at the MISTA Challenge (column MISTA), in addition to the improved post competition results obtained by Asta et al.[4] [1]. Average values were computed from five independent executions of the best *online* approach and five of the best *offline* approach.

Many new best known solution were obtained, in addition to improved average solution costs for many instances. considering Asta's results. Note that in [1],

---

[4] To the best of our knowledge the post competition results of Asta et al. [1] were not submitted to the MISTA Challenge official website.

Table 5. Comparative results of the LAHC metaheuristic

	LAHC				MISTA [9]		Asta et al. [1]			
	Best		Average		Best		Best		Average	
	TPD	TMS	TPD	TMS	TPD	TMS	TPD	TMS	TPD	TMS
A-1	**1**	**23**	**1**	**23**	1	23	1	23	1	23
A-2	**2**	**41**	**2**	**41**	2	41	2	41	2	41
A-3	**0**	**50**	**0**	**50**	0	50	0	50	0	50
A-4	**65**	**42**	66.2	43.0	65	42	65	42	65	42
A-5	153	104	163.0	108.6	153	105	**150**	**103**	155	105
A-6	**133**	**91**	147.4	96.0	147	96	133	99	**141**	**808**
A-7	597	203	625.0	202.6	596	196	**590**	**190**	605	201
A-8	**270**	**153**	**285.4**	**152.8**	302	155	272	148	292	153
A-9	**194**	**126**	**204.2**	**126.0**	223	119	197	122	208	128
A-10	845	308	**877.2**	**310.2**	969	314	**836**	**303**	880	313
B-1	352	128	365.4	130.6	349	127	**294**	**118**	**352**	**128**
B-2	431	162	**448.2**	**167.4**	434	160	**431**	**158**	452	167
B-3	530	208	**550.6**	**208.6**	545	210	**526**	**200**	554	210
B-4	1267	281	1313.6	284.4	1274	289	**1252**	**275**	**1299**	**283**
B-5	816	253	835.4	254.4	820	254	**807**	**245**	**832**	**255**
B-6	**877**	**222**	**917.6**	**226.0**	911	226	905	225	950	232
B-7	793	229	854.2	238.0	792	228	**782**	**225**	**802**	**232**
B-8	**2865**	**520**	**3065.2**	**530.0**	2974	541	2974	541	3323	545
B-9	**4059**	**743**	**4133.4**	**745.0**	4192	746	4062	738	4247	754
B-10	**3030**	**445**	**3094.0**	**442.4**	3050	448	3050	448	3290	455
X-1	390	143	427.6	145.6	392	142	**386**	**137**	**405**	**143**
X-2	351	163	366.6	164.6	349	163	**345**	**158**	**356**	**164**
X-3	**307**	**183**	**325.6**	**190.6**	324	192	310	187	329	193
X-4	909	206	**958.6**	**207.2**	915	208	**907**	**201**	960	209
X-5	1749	371	1786.4	374.0	1768	374	**1727**	**362**	**1785**	**373**
X-6	**686**	**226**	**708.4**	**230.2**	700	234	690	226	730	238
X-7	845	226	887.4	232.8	861	236	**831**	**220**	**866**	**233**
X-8	**1183**	**281**	**1235.2**	**282.4**	1218	286	1201	279	1256	288
X-9	**3084**	**626**	**3189.0**	**638.2**	3268	643	3155	632	3272	648
X-10	1580	381	1617.0	382.4	1600	381	**1573**	**383**	**1613**	**383**

the best known results were obtained by running 2,500 independent executions in a computer cluster.

All new best solutions were sent to the official website[5] of MISTA Challenge.

---

[5] https://gent.cs.kuleuven.be/mista2013challenge/.

8 Conclusions and Future Works

In this paper we evaluated different strategies to define the best neighborhood composition in stochastic local search methods. These strategies were evaluated in a metaheuristic with fourteen neighborhoods. The first important observation is how important the definition of appropriate neighborhood selection probabilities is: the trivial implementation with uniform neighborhood selection probabilities performed worse than versions with more elaborated strategies to define these values.

Our results provided some valuable insights on the role of each neighborhood. While some neighborhoods are quite useful in the beginning of the search, they may prove themselves ineffective when improved solutions are being processed. Other neighborhoods remain effective during the entire search. One future work would be to understand what makes these neighborhoods special and try to create new neighborhoods inspired in a way that they preserve these good characteristics or discard some neighborhoods that are not significant. Concerning the update of neighborhood selection probabilities during the search, for cases where strong dual bounds can be obtained during the search a *gap based* approach could be used to change neighborhood selection probabilities based on the current maximum relative distance to the optimal solution.

Online tuning approaches can be very useful to simplify the tuning process and to define the best neighborhood composition strategies during *runtime*. While this approach significantly improves the uniform probabilities strategies, it performs slight worse (1 % in our tests) than the best configuration obtained with *offline* tuning.

Future work include applying additional intensification and diversification strategies. Furthermore, the presented neighborhood composition may be employed in others metaheuristics.

Acknowledgments. The authors thank CNPq and FAPEMIG for supporting this research.

References

1. Asta, S., Karapetyan, D., Kheiri, H., Ozcan, E., Parkes, A.J.: Combining monte-carlo and hyper-heuristic methods for the multi-mode resource-constrained multi-project scheduling problem. In: Kendall, G., Berghe, G.V., McCollum, B. (eds.) Proceedings of the 6th Multidisciplinary International Scheduling Conference, (in review), pp. 836–839 (2013)
2. Baker, J.E.: Reducing bias and inefficiency in the selection algorithm. In: Proceedings of the 2nd International Conference on GA, pp. 14–21. Lawrence Erlbaum Associates, Inc., Mahwah, NJ, USA (1987)
3. da Fonseca, G.H.G., Santos, H.G., Toffolo, T.A.M., Brito, S.S., Souza, M.J.F.: GOAL solver: a hybrid local search based solver for high school timetabling. Ann. Oper. Res. **239**(1), 77–97 (2014)

4. Demeulemeester, E.L., Herroelen, W.S.: Project Scheduling: A Research Handbook. Kluwer Academic Publishers, Norwell (2002)
5. Hansen, P., Mladenović, N.: First vs. best improvement: an empirical study. Discrete Appl. Math. **154**(5), 802–817 (2006)
6. Hoos, H.H., Stützle, T.: Stochastic local search: Foundations and applications, pp. 149–201. Morgan Kaufmann (2005)
7. Ochoa, G., Verel, S., Daolio, F., Tomassini, M.: Local optima networks: a new model of combinatorial fitness landscapes. In: Richter, H., Engelbrecht, A. (eds.) Recent Advances in the Theory and Application of Fitness Landscapes. ECC, vol. 6, pp. 245–276. Springer, Heidelberg (2014)
8. Soares, J.A., Santos, H.G., Baltar, D., Toffolo, T.A.M.: LAHC applied to the multi-mode resource-constrained multi-project scheduling problem. In: Proceedings of 7th Multidisciplinary International Conference on Scheduling, pp. 905–908 (2015)
9. Wauters, T., Kinable, J., Smet, P., Vancroonenburg, W., Berghe, G.V., Verstichel, J.: The multi-mode resource-constrained multi-project scheduling problem:The MISTA 2013 challenge. J. Sched. (2014)

A Hybrid Multi-objective Evolutionary Approach for Optimal Path Planning of a Hexapod Robot

A Preliminary Study

Giuseppe Carbone[1,3(✉)] and Alessandro Di Nuovo[2,4]

[1] Department of Engineering and Mathematics,
Sheffield Hallam University, Sheffield, UK
g.carbone@shu.ac.uk
[2] Department of Computing, Sheffield Hallam University, Sheffield, UK
[3] Department of Civil and Mechanical Engineering,
University of Cassino and South Latium, Cassino, Italy
[4] Faculty of Engineering and Architecture, University of Enna "Kore", Enna, Italy

Abstract. Hexapod robots are six-legged robotic systems, which have been widely investigated in the literature for various applications including exploration, rescue, and surveillance. Designing hexapod robots requires to carefully considering a number of different aspects. One of the aspects that require careful design attention is the planning of leg trajectories. In particular, given the high demand for fast motion and high-energy autonomy it is important to identify proper leg operation paths that can minimize energy consumption while maximizing the velocity of the movements. In this frame, this paper presents a preliminary study on the application of a hybrid multi-objective optimization approach for the computer-aided optimal design of a legged robot. To assess the methodology, a kinematic and dynamic model of a leg of a hexapod robot is proposed as referring to the main design parameters of a leg. Optimal criteria have been identified for minimizing the energy consumption and efficiency as well as maximizing the walking speed and the size of obstacles that a leg can overtake. We evaluate the performance of the hybrid multi-objective evolutionary approach to explore the design space and provide a designer with an optimal setting of the parameters. Our simulations demonstrate the effectiveness of the hybrid approach by obtaining improved Pareto sets of trade-off solutions as compared with a standard evolutionary algorithm. Computational costs show an acceptable increase for an off-line path planner.

Keywords: Multi-objective optimization · Robot design · Legged robots · Hexapod robots

1 Introduction

Hexapod walking robots (HWR) are six-legged robots having a degree of autonomy that can range from partial autonomy, including teleoperation, to full autonomy without active human intervention [1, 2]. HWR usually have as high stability, low footprint,

© Springer International Publishing Switzerland 2016
M.J. Blesa et al. (Eds.): HM 2016, LNCS 9668, pp. 131–144, 2016.
DOI: 10.1007/978-3-319-39636-1_10

fault tolerant locomotion features [3]. They can also overcome obstacles that are comparable to the size of the robot leg [4]. These main characteristics make hexapod walking robots a suitable choice in several application scenarios such mine fields [5], planets exploration [6], search and rescue operations [7], forests harvesting [8]. Despite the above-referenced advantages and applications, many challenges remain in the field of hexapod locomotion. In fact, HWRs are still complex and slow machines, consisting of many actuators, sensors, transmissions and power supply hardware.

During the last years, the field of legged robots has been strongly influenced by the development of efficient optimization techniques, which coupled with low-cost and fast computational resources, have allowed for the resolution of such optimization problems. Nevertheless, challenges remain in the field of many-legged robot locomotion such as Hexapods. Hexapods are walking that are, in fact, complex and slowly machines, consisting of many actuators, sensors, transmissions and supporting hardware.

One of the approaches that need more investigation in the field of is multi-objective optimization (MOO), which involves minimizing or maximizing multiple objective functions subject to a set of constraints. Indeed, optimizing the design of a hexapod robot includes analysing and selecting design trade-offs between two or more conflicting objectives.

In the area of MOO, Multi-Objective Evolutionary Algorithms (MOEAs) demonstrated to be well-suited for solving several complex multi-objective problems [9, 10]. These algorithms adopt the same basic principles of the single-object evolutionary algorithm by emulating the evolutionary process on a set of individuals (solutions), i.e. an evolutionary population, by means of the so-called evolutionary operators (fitness assignment, selection, crossover, mutation, and elitism). In general, MOEAs differ on the fitness assignment method, but most of them are part of a family, called Pareto-based, which use the Pareto dominance concept as the foundation to discriminate solutions to guide their search [10]. For examples, the interested reader can refer to several surveys of multi-objective optimization methods, such as for engineering [11, 12], for data mining [13–15], for bio-informatics [16], for portfolio and other financial problems [17].

A previous attempt of using an MOEA to optimize the design of a leg mechanism has been presented in [18], where the authors compared the performance with that of an earlier study and in all cases the superiority and flexibility of the EMO approach was demonstrated. The MOEA used in this previous study was NSGA-II [19].

In this paper, we use a hybrid approach to extend and improve the previous result, which, at the best of our knowledge, is the only attempt of using an MOO approach to solve the problem of the design optimization of a robotic leg.

2 Materials and Methods

In this section, we briefly present the Materials and methods used in this work. For the brevity required by a conference paper, we are only presenting the main characteristics of the algorithms and of the robotic platform used. The interested reader should refer to the cited publications in the provided reference list for more details.

2.1   A Hybrid Multi-objective Evolutionary Approach

The MOEA we considered for our experiments is a controlled elitist genetic algorithm, which is a variant of the well know and widely used NSGA-II [19]. An elitist GA always favours individuals with better fitness value (rank) whereas, a controlled elitist GA also favours individuals that can help increase the diversity of the population even if they have a lower fitness value. In our application domain, it is very important to maintain the diversity of population for convergence to an optimal Pareto front. This is done by controlling the elite members of the population as the algorithm progresses. A non-dominated rank is assigned to each individual using the relative fitness. Individual a dominates b (a has a lower rank than b) if a is strictly better than b in at least one objective and a is no worse than b in all objectives. This is same as saying b is dominated by a or a is non-inferior to b. Two individuals a and b are considered to have equal ranks if neither dominates the other. The distance measure of an individual is used to compare individuals with equal rank. It is a measure of how far an individual is from the other individuals with the same rank. For the rest, the standard process of evolutionary algorithms still applies. It works on a population using a set of operators that are applied to the population. A population is a set of points in the design space. The initial population is generated randomly by default. The next generation of the population is computed using the non-dominated rank and a distance measure of the individuals in the current generation.

To increase the performance of the MOEA we used a hybrid scheme to find an optimal Pareto front for our MO problem. In fact, an MOEA can reach the region near an optimal Pareto front relatively quickly, but it can take many further function evaluations to achieve convergence. For this reason, a commonly used technique is to run the MOEA for a relatively small number of generations to get near an optimum front. Then the Pareto set solution obtained by the MOEA is used as an initial point for another optimization solver that is faster and more efficient for a local search. We used the Goal Attainment Method [20] as the hybrid solver, which reduces the values of a linear or nonlinear vector function to attain the goal values given in a goal vector. The method used is a sequential quadratic programming (SQP), which represent the state of the art in nonlinear programming methods [21]. The slack variable γ is used as a dummy argument to minimize the vector of objectives simultaneously; the goal is a set of values that the objectives attain. In our case, the goals were set as 0, while the starting point was the Pareto set obtained by the MOEA.

In our experiments, we used the MATLAB 2015a implementation for both algorithms, further details can be found in the software documentation.

2.2   Measures for Comparing the Quality of the Results

The main performance measure we considered is the *hypervolume* [22], that is the only one widely accepted and, thus, used in many recent similar works. This index measures the hypervolume of that portion of the objective space that is weakly dominated by the Pareto set to be evaluated. The estimation is done through 10^6 uniformly distributed random points within the bounded rectangle. We took as bounding point vector [1000, 100], because these are the maximum realistic values we allowed for the design of the hexapod robot [23].

Pareto *dominance* is equal to the ratio between the total number of points in Pareto-set P that are also present in a reference Pareto-set R (i.e., it is the number of non-dominated points by the other Pareto-set). In this case, a higher value obviously corresponds to a better Pareto-set. Using the same reference Pareto-set, it is possible to compare quantitatively results from different algorithms.

The reference Pareto was obtained in the following way: first, we combined all approximations sets generated by the algorithms under consideration, and then the dominated objective vectors are removed from this union. At last, the remaining points, which are not dominated by any of the approximations sets, form the reference set. The advantage of this approach is that the reference set weakly dominates all approximation sets under consideration [12, 24].

We also calculated the *computational efficiency* calculated as the total time spent by the MATLAB routines for each run of one approach divided by the number of generations for that run. This has been preferred to the simple computation time because for each run of the MOEA a different number of generations were employed by the algorithm for obtaining the Pareto set. The computational efficiency allows a direct comparison of all the runs. The tests have been done on an Intel® Core™ i7-3770 3·40 GHz using 4 parallel threads.

2.3   The Cassino Hexapod Robot

In a recent past, research activities have been undergoing at LARM, Laboratory of Robotics and Mechatronics of Cassino and Southern Lazio University, for developing six-legged robots within the so-called "Cassino Hexapod" series (for more details see [25–28]. Fig. 1 shows a prototype of Cassino Hexapod II. The main features of the proposed design solutions have been the use of low-cost mechanism architectures and user-friendly operation features. Cassino Hexapod is legged walking robot, whose intended main application task is the inspection and analysis of historical sites. In particular, the robot should be able to move inside archeological and/or architectural

Fig. 1. The Cassino Hexapod II

sites by carrying surveying devices and by avoiding damage to the delicate surfaces or historical items of the site. Additionally, the robot should be able to operate also in environments that cannot be reached or that are unsafe for human operators.

3   Kinematic Model of One Leg

The kinematic model of one leg can be established by considering two links in a 3R configuration as shown in Fig 2. The 3 R revolute joints have parallel rotation axes. The first and second revolute joints are connected to the first and second link, respectively. The third revolute joint is allowing the rotation of a wheel relative to the second link. The kinematic path planning task consists of identifying proper values of the revolute joint angles θ_1 and θ_2 as a function of time. Typically, a robot controller will update the values of the joint angles θ_1 and θ_2 at a fixed clock speed rate that can be assumed as equal to 10 ms. Values of joint angles are often obtained in path planning techniques by search algorithms or by means of interpolation equations such as 5th order polynomials, as proposed for example by Frankovský et al. [29]. Accordingly, for the joint angles $\theta 1$ and $\theta 2$ one can write

$$\theta_1(t) = a_1 t^5 + a_2 t^4 + a_3 t^3 + a_4 t^2 + a_5 t + a_6 \tag{1}$$

$$\theta_2(t) = b_1 t^5 + b_2 t^4 + b_3 t^3 + b_4 t^2 + b_5 t + b_6 \tag{2}$$

Specific boundary conditions can help in simplifying the required models by reducing the number of parameters to be searched or set-up in Eqs. (1) and (2). For example, one can assume that a leg motion starts from the fully straight leg configuration having θ_1 and θ_2 equal to zero. Additionally, the initial and final angular speed and acceleration can be assumed as equal to zero at the beginning and at the end of a leg motion. Based on the above boundary conditions one can set up the following parameters in Eqs. (1) and (2)

$$\theta_1(t = 0) = 0 \rightarrow a_6 = 0$$
$$\theta_2(t = 0) = 0 \rightarrow b_6 = 0$$
$$\dot{\theta}_1(t = 0) = 0 \rightarrow a_5 = 0$$
$$\dot{\theta}_2(t = 0) = 0 \rightarrow b_5 = 0$$
$$\ddot{\theta}_1(t = 0) = 0 \rightarrow a_4 = 0$$
$$\ddot{\theta}_2(t = 0) = 0 \rightarrow b_4 = 0$$

Accordingly, the kinematic path planning of a leg requires identifying the parameters in Eqs. (1) and (2). These parameters can be obtained by using search scripts in optimization algorithms.

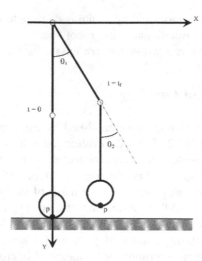

Fig. 2. Kinematic scheme of a robotic leg

4 Dynamic Model of One Leg

Dynamic effects play a significant role in the operation of a leg especially as referring to energy consumption and operation speeds. Accordingly, a basic dynamic model has been established by referring to the basic double pendulum architecture of a leg as shown in (Fig. 3). Accordingly, dynamic equations can be established by referring to the Euler-Lagrange formulation in the form

$$\frac{d}{dt}\left(\frac{\partial L}{\partial \dot{q}_i}\right) - \frac{\partial L}{\partial q_i} = \tau_i \tag{3}$$

in which
  $i = 1, 2 \ldots n$
  L = Lagrangian = $T - U$
  T = total kinetic energy of the system
  U = Potential energy of the system
  q_i = generalized coordinates of the manipulator
  $\dot{q}_i$ = time derivatives of the generalized coordinates
  τ_i = generalized force (torque) that is needed at the i-th joint for moving the link l_i
  The inverse dynamic problem can be written by referring to Eq. (3) in terms of the torques τ_i that are needed to obtain the prescribed movement of the leg. Inputs are the prescribed θ_1 and θ_2 versus time as obtained from Eqs. (1) and (2).
  Using the above-mentioned values of θ_1 and θ_2 and referring to the model in Fig. 3 one can calculate the coordinates of the leg joints in the form

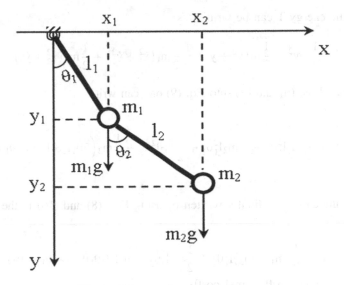

Fig. 3. Scheme of control architecture

$$x_1 = l_1 \text{sen}\theta_1; \ y_1 = -l_1 \cos\theta_1;$$
$$x_2 = l_1 \text{sen}\theta_1 + l_2 \text{sen}\theta_2; \ y_2 = -l_1 \cos\theta_1 - l_2 \cos\theta_2 \tag{4}$$

The time derivatives of Eq. (4) can be written as

$$\dot{x}_1 = l_1 \cos\theta_1 \cdot \dot{\theta}_1 \qquad\qquad \dot{y}_1 = l_1 \text{sen}\theta_1 \cdot \dot{\theta}_1$$
$$\dot{x}_2 = l_1 \cos\theta_1 \cdot \dot{\theta}_1 + l_2 \cos\theta_2 \cdot \dot{\theta}_2 \ \dot{y}_2 = l_1 \text{sen}\theta_1 \cdot \dot{\theta}_1 + l_2 \text{sen}\theta_2 \cdot \dot{\theta}_2 \tag{5}$$

Equation (5) can be also used to write

$$\dot{x}_1^2 + \dot{y}_1^2 = l_1^2 \cos^2\theta_1 \cdot \dot{\theta}_1^2 + l_1^2 \text{sen}^2\theta_1 \cdot \dot{\theta}_1^2 = l_1^2 \cdot \dot{\theta}_1^2 \tag{6}$$

Equation (5) can be also rewritten as follows

$$\dot{x}_2^2 = l_1^2 \cos^2\theta_1 \cdot \dot{\theta}_1^2 + l_2^2 \cos^2\theta_2 \cdot \dot{\theta}_2^2 + 2l_1 l_2 \cos\theta_1 \cos\theta_2 \dot{\theta}_1 \dot{\theta}_2$$
$$\dot{y}_2^2 = l_1^2 \text{sen}^2\theta_1 \cdot \dot{\theta}_1^2 + l_2^2 \text{sen}^2\theta_2 \cdot \dot{\theta}_2^2 + 2l_1 l_2 \text{sen}\theta_1 \text{sen}\theta_2 \dot{\theta}_1 \dot{\theta}_2 \tag{7}$$
$$\dot{x}_2^2 + \dot{y}_2^2 = l_1^2 \cdot \dot{\theta}_1^2 + l_2^2 \cdot \dot{\theta}_2^2 + 2l_1 l_2 \dot{\theta}_1 \dot{\theta}_2 \cos(\theta_1 - \theta_2)$$

Considering the effects of gravity, in terms of mass and inertia in Eqs. (4)–(6) one can write the potential energy U as

$$U = m_1 g y_1 + m_2 g y_2 = -m_1 g l_1 \cos\theta_1 + m_2 g(l_1 \cos\theta_1 + l_2 \cos\theta_2) \tag{8}$$

The kinetic energy T can be written as

$$T = \frac{1}{2}mv^2 = \frac{1}{2}m(\dot{x}^2 + \dot{y}^2) = \frac{1}{2}m_1(\dot{x}_1^2 + \dot{y}_1^2) + \frac{1}{2}m_2(\dot{x}_2^2 + \dot{y}_2^2) \tag{9}$$

Substituting Eqs. (6) and (7) into Eq. (9) one can write

$$T = T_1 + T_2 = \frac{1}{2}m_1l_1^2\dot{\theta}_1^2 + \frac{1}{2}m_2l_1^2\dot{\theta}_1^2 + \frac{1}{2}m_2l_2^2\dot{\theta}_2^2 + \frac{1}{2}m_2\left(2\dot{\theta}_1l_1\dot{\theta}_2l_2\cos(\theta_1 - \theta_2)\right) \tag{10}$$

The Lagrangian can be finally written by using Eqs. (8) and (10) in the form

$$L = T - U = \frac{1}{2}(m_1 + m_2)l_1^2\dot{\theta}_1^2 + \frac{1}{2}m_2l_2^2\dot{\theta}_2^2 + m_2l_1l_2\dot{\theta}_1\dot{\theta}_2\cos(\theta_1 - \theta_2)$$
$$+ (m_1 + m_2)gl_1\cos\theta_1 + m_2l_2\cos\theta_2 \tag{11}$$

Substituting Eq. (11) in Eq. (3) leads to the calculation of the required input torques τ_i that are needed to obtain the prescribed movement of the leg in the form

$$\tau_1 = (m_1 + m_2)l_1^2\ddot{\theta}_1 + m_2l_1l_2\ddot{\theta}_2\cos(\theta_1 - \theta_2) + m_2l_1l_2\dot{\theta}_2^2\mathrm{sen}(\theta_1 - \theta_2) + gl_1(m_1 + m_2)\mathrm{sen}\theta_1$$
$$\tau_2 = m_2l_2^2\ddot{\theta}_2 + m_2l_1l_2\ddot{\theta}_1\cos(\theta_1 - \theta_2) - m_2l_1l_2\dot{\theta}_1^2\mathrm{sen}(\theta_1 - \theta_2) + l_2m_2g\,\mathrm{sen}\theta_2 \tag{12}$$

A formulation for optimal path planning problem

The path planning task for a hexapod leg with n DoFs can be described using m knots in the trajectory of each k-th joint of a manipulator. The prescribed task can be given by the initial and final points P_0 and P_m of the trajectory. The movement of the leg can be obtained by the simultaneous motion of the n joints in order to perform the prescribed task. Among the many available criteria, one can assume the energy aspect as one of the most significant performance in order to optimize the manipulator operation, since the energy formulation can consider simultaneously dynamic and kinematic characteristics of the performing motion. It should also be considered that a maximization of the operation speed of a leg corresponds to a maximization of the amplitude of the movement, when time is fixed.

An optimality criterion concerning with energy aspects of the path motion can be conveniently expressed in terms of the work that is needed by the actuators. In particular, the work by the actuators is needed for increasing the kinetic energy of the system in a first phase from a rest condition to actuators states at which each actuator is running at maximum velocity. In a second phase bringing the system back to a rest condition, the kinetic energy will be decreased to zero through the actions of actuators and brakes. The potential energy of the system will contribute to size the necessary work by the actuators and friction effects in the joints can be assumed as negligible as compared to the actions of actuators and brakes. Thus, we have considered convenient to use the work W_{act} done by the actuators in the first phase of the path motion as an optimality criterion for optimal path generation as given by the expression

$$W_{act} = \sum_{k=1}^{3} \left[\int_{0}^{t_k} \tau_k \, \dot{\alpha}_k \, dt \right] \tag{13}$$

in which τ_k is the k-th actuator torque; $\dot{\alpha}_k$ is the k-th shaft angular velocity of the actuator, and t_k is the time coordinate value delimiting the first phase of path motion with increasing speed of the k th actuator.

Therefore, minimizing W_{act} has the aim to size at the minimum level the design dimensions and operation actions of the actuators while generating a path between two given extreme positions. Indeed, in general, once the actuator work is minimized, energy consumption of the system operation will be optimized consequently.

The other factor to optimize is the average speed, which is as follows:

$$v_{avg} = \frac{x_2 \quad x_1}{t_f - t_0} \tag{14}$$

in which x_1 and x_2 are the coordinates at the beginning (t_0) and at the end (t_f) of the movement. Note that, in this work, we assumed $(y_2 \quad y_1) = (x_2 \quad x_1)$ due to the symmetry of the model that has been proposed in Fig. 3.

The two optimal criteria that have been considered are conflicting, because an increase in the speed will result in higher energy consumption. Therefore, it can be established a multi-criteria optimization problem as follows:

$$min_{\mathbf{d}}(W_{act}(\mathbf{d}), \frac{1}{v_{avg}(\mathbf{d})}) \tag{15}$$

where $\mathbf{d}$ is the vector of design variables: $[a_1 \; a_2 \; a_3 \; b_1 \; b_2 \; b_3]$.

5 Experimental Results

Using the experimental setup described in Sect. 2, we ran both standard MOEA and the hybrid MOEA twenty-one times with different random seed. Identical parameter setting is used for the MOEAs: population size = 300; Tournament selection, size = 4; number of individuals in the Pareto set = 100; Elite count = 15; individual recombination (crossover) probability = 0·8; Gaussian mutation function. The MOEA stopped when the average change in the spread of the Pareto front over last 100 generations is less than 10^{-4}.

Median values for performance indicators are presented to represent the expected (mid-range) performance. For the analysis of multiple runs, we compute the quality measures of each individual run, and report the median and the standard deviation of these. Since the distribution of the algorithms we compare are not necessarily normal, we use the Mann-Whitney U test (a.k.a. Wilcoxon rank sum test) test [30] to indicate if there is a statistically significant difference between distributions. We recall that the significance level of a test is the maximum probability p, assuming the null hypothesis, which the statistic will be observed, i.e. the null hypothesis will be rejected in error when it is true. The lower the significance level the stronger the evidence. In this work, we assume that the null hypothesis is rejected if $p < 0.01$.

Considering all the runs, the median number of generations was 229 (standard deviation = 102, minimum = 10; maximum = 544).

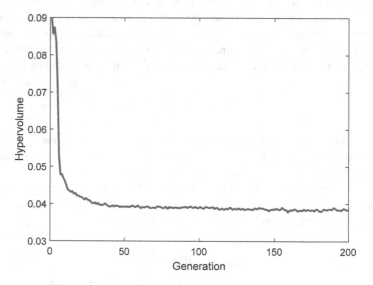

Fig. 4. Average hypervolume values over generations.

Figure 4 shows the hypervolume values over generations. We can see that the stop condition (Pareto spread is lower than 10–4) still allows a quite high number of generations without significant improvements in terms of minimization of the hypervolume. This test confirms that we should not expect a significant increase in the performance with more generations.

Table 1 presents the experimental results on the three performance measures considered. For all measures, the Hybrid MOEA significantly outperforms the Standard MOEA. This is evidenced in Fig. 5, which reports the box plots for hypervolumes and efficiency comparison. It is evident from the figure the significant increase given by the hybrid approach used at the average cost of 0·1422 s for a generation. This is confirmed by the statistical test, which rejects the hypotheses that the distributions are the same in both cases.

Table 1. Multi-metric comparison of the Pareto sets obtained by the standard and hybrid MOEAs. Values are the medians of each distribution and the standard deviation is in parentheses. Statistical significance (p) has been evaluated with the Mann-Whitney U test (hypervolume and efficiency: the lower the better; reference and dominance: the higher the better).

Measure	Standard MOEA	Hybrid MOEA	p
Hypervolume	*0·0367(0·0013)*	***0·0343 (0·0014)***	<0·001
Dominance (%)	*0·00 (0021)*	***4·39 (1·18)***	<0·001
Efficiency (sec/gen)	***0·2301 (0·0860)***	*0·3723 (0·0851)*	<0·001

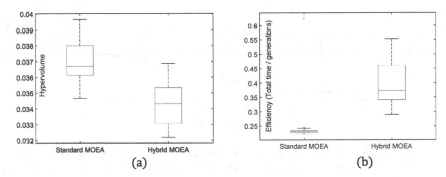

Fig. 5. Box plot comparisons: (a) hypervolumes of 21 runs of the standard and hybrid MOEA; (b) computational efficiency calculated as the total time divided by the number of generations (variable for each run).

Figure 6 presents the cumulative Pareto Sets obtained merging the Paretos of each run of the two approaches. These graphically confirm the numerical results that the hybrid approach significantly increases the performance of the MOEA. In particular, comparing the two cumulative Pareto sets, we can see that the improvement is well spread along the objective space and the most significant results is achieved in the central area which is the most common selection for the designer.

Indeed, from the designer point of view, the reader can clearly see the improvement given by the hybrid approach over the standard MOEA can be seen in Fig. 7, which plots the length and height of the movement over energy. Indeed, the possible designs obtained by the Hybrid MOEA can produce a movement of 0·3520 m amplitude (length and height) consuming only 89·60 W, while the matching solution of the standard MOEA requires 286·21 W (3 times more) for the a similar amplitude (0·3502).

In terms of real applicability of the solutions, this result can allow to include smaller batteries and, thus, increase the available payload. Furthermore, one can note that the leg can reach about 3·5 times the link length (not considering the wheel radius) with energy values that can be even lower than 100 W. These values can be seen as feasible also as compared to standards [1].

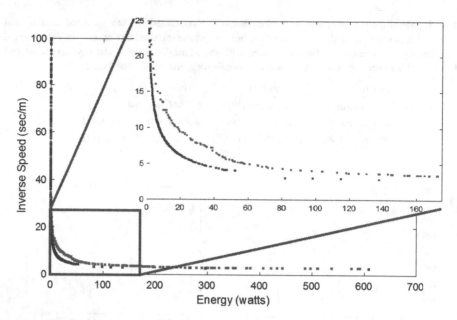

Fig. 6. Best cumulative Pareto sets comparison: standard MOEA Pareto set (red) and the hybrid MOEA (blue). The central area of the figure is zoomed to highlight the difference (Color figure online).

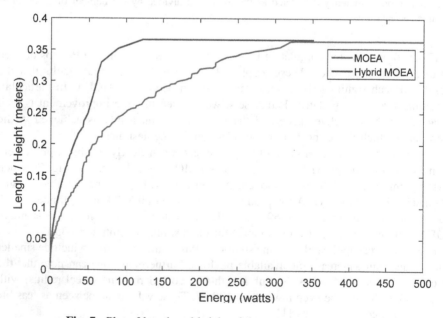

Fig. 7. Plot of length and height of the movement over energy.

6 Conclusions

In this paper, we present a preliminary study on the use of a hybrid multi-objective evolutionary approach for the optimal path planning of a hexapod robot leg. To this end, we evaluated the performance of a hybrid multi-objective optimization approach to explore the design space and provide the designer with the optimal setting of the parameters. To preliminary assess the optimization approach, a kinematic and dynamic model of a leg of a hexapod robot has been proposed as referring to the main design parameters. Optimal criteria have been identified for minimizing the energy consumption and efficiency as well as maximizing the size of obstacles that the robot can overtake. In our simulations, the hybrid approach demonstrated to achieve statistically significantly better Pareto sets of trade-off solutions than the standard evolutionary algorithm with acceptable time increase. These solutions are also better in comparison with other non-evolutionary algorithms applied to similar design problems. Our future work will focus on the application of the hybrid MOEA approach to the optimized design of all the six legs of the robot, which is a constrained optimization problem with a larger design space to explore.

References

1. Tedeschi, F., Carbone, G.: Design issues for hexapod walking robots. Robotics **3**, 181–206 (2014)
2. International Federation of Robotics. http://www.ifr.org/
3. Chavez-Clemente, D.: Gait optimization for multi-legged walking robots, with application to a lunar hexapod (2011). http://purl.stanford.edu/px063cb7934Includes
4. Carbone, G., Ceccarelli, M.: A mechanical design of a low-cost easy-operation anthropomorphic wheeled leg for walking machines. Int. J. Robot. Manag. **9**, 3–8 (2004)
5. De Santos, P.G., Garcia, E., Estremera, J.: Quadrupedal locomotion: an introduction to the control of four-legged robots. Springer, London (2006)
6. Krotkov, E.P., Simmons, R.G., Whittaker, W.L.: Ambler: performance of a six-legged planetary rover. Acta Astronaut. **35**, 75–81 (1995)
7. Nonami, K., Barai, R.K., Irawan, A., Daud, M.R.: Hydraulically Actuated Hexapod Robots: Design, Implementation and Control. Intelligent Systems, Control and Automation: Science and Engineering, vol. 66. Springer, Heidelberg (2014)
8. Silva, M.F., Machado, J.A.T., Barbosa, R.S.: Complex-order dynamics in hexapod locomotion. Sig. Process. **86**, 2785–2793 (2006)
9. Coello, C.A.C., Lamont, G.B., Van Veldhuizen, D.A: Evolutionary Algorithms for Solving Multi-objective Problems, 2nd edn. Springer, Heidelberg (2007)
10. Deb, K.: Multi-objective Optimization Using Evolutionary Algorithms. Wiley, New York (2001)
11. Marler, R.T., Arora, J.S.: Survey of multi-objective optimization methods for engineering. Struct. Multidiscip. Optim. **26**, 369–395 (2004)
12. Ascia, G., Catania, V., Di Nuovo, A.G., Palesi, M., Patti, D.: Performance evaluation of efficient multi-objective evolutionary algorithms for design space exploration of embedded computer systems. Appl. Soft Comput. **11**, 382–398 (2011)

13. Mukhopadhyay, A., Maulik, U., Bandyopadhyay, S., Coello, C.A.C.: A survey of multiobjective evolutionary algorithms for data mining: part i. IEEE Trans. Evol. Comput. **18**, 4–19 (2014)
14. Mukhopadhyay, A., Maulik, U., Bandyopadhyay, S., Coello, C.A.C.: Survey of multiobjective evolutionary algorithms for data mining: part II. IEEE Trans. Evol. Comput. **18**(1), 20–35 (2014)
15. Di Nuovo, A., Palesi, M., Catania, V.: Multi-objective evolutionary fuzzy clustering for high-dimensional problems. In: IEEE International Conference on Fuzzy Systems, FUZZ-IEEE 2007, pp. 1–6. IEEE (2007)
16. Di Nuovo, A., Catania, V.: An evolutionary fuzzy c-means approach for clustering of bio-informatics databases. In: IEEE International Conference on Fuzzy Systems, FUZZ-IEEE 2008. (IEEE World Congress on Computational Intelligence), pp. 2077–2082. IEEE (2008)
17. Ponsich, A., Jaimes, A.L., Coello, C.A.C.: A survey on multiobjective evolutionary algorithms for the solution of the portfolio optimization problem and other finance and economics applications. IEEE Trans. Evol. Comput. **17**, 321–344 (2013)
18. Deb, K., Tiwari, S.: Multi-objective optimization of a leg mechanism using genetic algorithms. Eng. Optim. **37**, 325–350 (2005)
19. Deb, K., Pratap, A., Agarwal, S., Meyarivan, T.: A fast and elitist multiobjective genetic algorithm: NSGA-II. IEEE Trans. Evol. Comput. **6**, 182–197 (2002)
20. Fleming, P.J., Pashkevich, A.P.: Application of multi-objective optimisation to compensator design for SISO control systems. Electron. Lett. **22**(1), 258–259 (1986)
21. Schittkowski, K.: NLPQL: A fortran subroutine solving constrained nonlinear programming problems. Ann. Oper. Res. **5**, 485–500 (1986)
22. Zitzler, E., Thiele, L., Laumanns, M., Fonseca, C.M., Da Fonseca, V.G.: Performance assessment of multiobjective optimizers: an analysis and review. IEEE Trans. Evol. Comput. **7**(2), 117–132 (2003)
23. Tedeschi, F., Carbone, G.: Design of hexapod walking robots: background and challenges. In: Habib, M.K. (ed.) Handbook of Research on Advancements in Robotics and Mechatronics, pp. 527–566. IGI Global (2015)
24. Di Nuovo, A.G., Ascia, G., Catania, V.: A study on evolutionary multi-objective optimization with fuzzy approximation for computational expensive problems. In: Coello, C.A., Cutello, V., Deb, K., Forrest, S., Nicosia, G., Pavone, M. (eds.) PPSN 2012, Part II. LNCS, vol. 7492, pp. 102–111. Springer, Heidelberg (2012)
25. Carbone, G., Shrot, A., Ceccarelli, M.: Operation strategy for a low-cost easy operation Cassino Hexapod. Appl. Bionics Biomech. **4**, 149–156 (2007)
26. Carbone, G., Tedeschi, F.: A low cost control architecture for Cassino Hexapod II. Int. J. Mech. Control. **14**, 19–24 (2013)
27. Carbone, G., Ceccarelli, M., Oliveira, P., Saramago, S.: Carvalho, J, F.: An optimum path planning for Cassino parallel manipulator by using inverse dynamics. Robotica **6**, 229–239 (2008)
28. Carbone, G., Tedeschi, F., Gallozzi, A., Cigol, M.: A robotic mobile platform for service tasks in cultural heritage. Int. J. Adv. Robot. Syst. **12**, 1–10 (2015)
29. Frankovský, P., Hroncová, D., Delyová, I., Hudák, P.: Inverse and forward dynamic analysis of two link manipulator. In: Procedia Engineering, pp. 158–163 (2012)
30. Mann, H.B., Whitney, D.R.: On a test of whether one of two random variables is stochastically larger than the other. Ann. Math. Stat. **18**, 50–60 (1947)

Hybridization of Chaotic Systems and Success-History Based Adaptive Differential Evolution

Adam Viktorin[(✉)], Roman Senkerik, and Michal Pluhacek

Faculty of Applied Informatics, Tomas Bata University in Zlin,
T. G. Masaryka 5555, 760 01 Zlin, Czech Republic
{aviktorin, senkerik, pluhacek}@fai.utb.cz

Abstract. This research paper focuses on hybridization of two soft computing fields – chaos theory and evolutionary algorithms, specifically on the implementation of Chaotic map based Pseudo-Random Number Generator (CPRNG) into the process of parent selection in Success-History Based Adaptive Differential Evolution (SHADE) algorithm. The impact on performance of the algorithm is tested on CEC2015 benchmark set where five different chaotic maps are used for random integer generation. Performance comparison shows that there is a potential in replacing classic Pseudo-Random Number Generators (PRNGs) with chaotic ones. The results provided in this paper show that the choice of CPRNG for given problem is crucial in terms of affecting the performance of the algorithm, therefore the next research step will be focused on the development of the framework which will adapt to the solved problem and select the most suitable CPRNG or their combination.

Keywords: Success-history based adaptive differential evolution · Deterministic chaos · Optimization · Parent selection · Pseudo-random number generator

1 Introduction

Since its introduction in 1995 [1], Differential Evolution (DE) was proved to be a simple yet effective heuristic method for solving optimization problems over continuous spaces [2–5].

The canonical version of DE has four control parameters; population size NP, number of generations G, scaling factor F and crossover rate CR. The performance of the DE algorithm depends on the values of these four control parameters and since the optimal setting of the control parameters varies for different optimization problems, a lot of research has been done in order to produce a strategy which would adapt control parameter values to given objective function. Some of the existing algorithms which adapt control parameters during evolution are jDE [6], SDE [7], SaDE [8] and JADE [9]. The latter also implements mutation strategy "current-to-$p$best/1" and an optional archive of inferior solutions A. JADE algorithm formed a base for a Success-History based Adaptive Differential Evolution (SHADE) [10], which extends JADE with two

© Springer International Publishing Switzerland 2016
M.J. Blesa et al. (Eds.): HM 2016, LNCS 9668, pp. 145–156, 2016.
DOI: 10.1007/978-3-319-39636-1_11

historical memories M_{CR}, M_F. These store sets of historically successful CR and F values respectively.

Recent research implies that the utilization of Chaotic map based Pseudo-Random Number Generators (CPRNGs) is promising for various implementations in Evolutionary Algorithms (EAs) and Swarm Intelligence (SI) algorithms [11–15].

This paper presents the results of five different CPRNGs used for parent selection in the SHADE algorithm and compares them to the canonical version of SHADE algorithm on the CEC2015 benchmark set [16]. Since the results suggest, that the performance might be improved when using CPRNGs for parent selection, the future research will focus on the development of a framework that would adaptively select the most suitable CPRNG for a given problem.

The paper is structured as follows: Next section focuses on CPRNGs. Section three briefly describes the SHADE algorithm and implementation of a CPRNG into the process of parent selection. Experiment setting and results are covered in sections four and five respectively. Sections that follow are result discussion and conclusion.

2 Chaotic Map Based PRNGs

Chaotic maps are systems generated from a single initial position by simple equations. The current position is used for the generation of a new position, thus creating a sequence that is extremely sensitive to the initial position, which is known as the "butterfly effect." In this research, Pseudo-Random Number Generator (PRNG) with uniform distribution was used for the generation of initial positions of five different discrete chaotic maps. Each of these chaotic systems has their own equations which are given in Table 1 together with their parameter settings. These were set according to [17].

Table 1. Chaotic systems and their parameter values.

Chaotic map	Equations	Parameters	Initial position
Burgers	$X_{n+1} = aX_n - Y_n^2$	$a = 0.75$	$X_0 = U[-0.1, -0.01]$
	$Y_{n+1} = bY_n + X_nY_n$	$b = 1.75$	$Y_0 = U[0.01, 0.1]$
Delayed logistic	$X_{n+1} = AX_n(1 - Y_n)$	$A = 2.27$	$X_0 = Y_0 = U[0.8, 0.9]$
	$Y_{n+1} = X_n$		
Dissipative	$X_{n+1} = X_n + Y_{n+1}(\text{mod } 2\pi)$	$b = 0.1$	$X_0 = Y_0 = U[0, 0.1]$
	$Y_{n+1} = bY_n + k \sin X_n(\text{mod } 2\pi)$	$k = 8.8$	
Lozi	$X_{n+1} = 1 - a\|X_n\| - bY_n$	$a = 1.7$	$X_0 = Y_0 = U[0, 0.1]$
	$Y_{n+1} = X_n$	$b = 0.5$	
Tinkerbell	$X_{n+1} = X_n + Y_n + aX_n + bY_n$	$a = 0.9$	$X_0 = U[-0.1, -0.01]$
		$b = -0.6$	
	$Y_{n+1} = 2X_nY_n + cX_n + dY_n$	$c = 2$	$Y_0 = U[0, 0.1]$
		$d = 0.5$	

The equation depicted in (1) was used to transform chaotic map generated sequence into a pseudo-random sequence of integer values from range [1, *maxRndInt*].

$$rndInt_i = \text{round}\left(\frac{\text{abs}(X_i)}{\text{max}(\text{abs}(X_{i \in N}))} * (maxRndInt - 1)\right) + 1 \qquad (1)$$

Where $\text{max}(\text{abs}(X_{i \in N}))$ is a maximum of absolute values of X from chaotic sequence of size N, which was generated by chaotic map.

3   Success-History Based Adaptive Differential Evolution and Chaos Induced Parent Selection

Since the introduction of DE in 1995 [1] the algorithm itself was improved in many ways. The canonical version of DE [1] can be divided into simple steps which are repeated in each iteration of the evolutionary process. Before the evolutionary process begins, there is the initialization phase in which the control parameters are set (population size NP, number of generations G, scaling factor F and crossover rate CR) and the first generation of candidate solutions (individuals) is generated randomly from the objective space. Iteration steps of canonical DE are mutation ("rand/1/bin" mutation strategy with random selection of parent vectors and static F), crossover (with static CR value) and elitism (only better solutions may be placed into the new generation).

The convergence speed and ability to reach the global optimum were the main targets for improvements of the original algorithm. Novel approaches to DE were tested on benchmark sets in order to compare them. One of the best to date state-of-art DE algorithm variants is SHADE.

The SHADE algorithm is based on JADE which adapts the values of F and CR continuously after each generation, it uses novel mutation strategy "current-to-*p*best/1" and implements external archive A. This archive preserves parent vectors which were worse than trial vectors in the elitism phase. SHADEs improvement to JADE is an implementation of two historical memories for successful F and CR values, M_F and M_{CR}. These memories maintain a diverse set of parameters to guide control parameter adaptation as search progresses. Even after all modifications, the SHADE algorithm still maintains simplicity of the canonical DE [1]. Concepts of basic operations in SHADE are displayed in next sections. For a detailed description of feature constrain correction, historical memories updates and external archive handling see [10].

3.1   Initialization

The initial population is generated randomly from objective space and has NP individuals. The content of historical memories M_F and M_{CR} of size H is initialized to be $M_{F,i} = M_{CR,i} = 0.5$ for ($i = 1, ..., H$). External archive A is empty and the number of generations G is defined by user.

3.2   Mutation Strategy and Parent Selection

As aforementioned, the SHADE algorithm uses "current-to-$p$best/1" mutation strategy depicted in (2) which adjusts greediness by parameter p ($p = U[p_{min}, 0.2]$, $p_{min} = 2/NP$).

$$v_{i,G} = x_{i,G} + F_i(x_{pbest,G} - x_{i,G}) + F_i(x_{r1,G} - x_{r2,G}) \qquad (2)$$

The individual $x_{i,G}$ is a i-th individual from members of the current generation G = population P, $x_{pbest,G}$ is selected from $NP \times p$ best individuals in P by $rndInt$ from CPRNG sequence (1). Individual $x_{r1,G}$ is randomly selected from P by $rndInt$ from CPRNG sequence and individual $x_{r2,\,G}$ is randomly selected from the union of population P and external archive A by $rndInt$ from CPRNG sequence, where $i \neq r1 \neq r2$. Scaling factor F_i is from a Cauchy distribution with the location parameter value of $M_{F,r}$ which is a randomly selected value from historical memory M_F and scale parameter value of 0.1 (3).

$$F_i = C[M_{F,r}, 0.1] \qquad (3)$$

3.3   Crossover and Elitism

The crossover rate for i-th individual CR_i is generated similarly to scaling factor F_i with a help of normal distribution and historical memory M_{CR} (4). Crossover operation is binomial as in canonical DE, where at least one feature has to be selected from vector v_i. This feature is determined by the random integer j_{rand} (5). The individual which survives to the next generation is selected from a pair of original x_i and trial u_i vectors based on their fitness values (6).

$$CR_i = N[M_{CR,r}, 0.1] \qquad (4)$$

$$u_{j,i,G} = \begin{cases} v_{j,i,G} & \text{if } rnd[0,1] \leq CR_i \text{ or } j = j_{rand} \\ x_{j,i,G} & \text{otherwise} \end{cases} \qquad (5)$$

$$x_{i,G+1} = \begin{cases} u_{i,G} & \text{if } f(u_{i,G}) < f(x_{i,G}) \\ x_{i,G} & \text{otherwise} \end{cases} \qquad (6)$$

If the trial vector u_i has better fitness value than the original vector x_i, scaling factor F_i and crossover rate CR_i are stored into corresponding temporary sets S_F and S_{CR} which serve as a base for evaluation of new values in historical memories after each generation. Also the original vector x_i is stored in an external archive A.

4 Experiment Setting

The canonical SHADE as well as SHADE with chaos induced parent selection were evaluated on CEC2015 benchmark set functions in 10 dimensions [16]. Each algorithm had the same setting of control parameters in compliance with [10]. CPRNGs were set up with initial parameters generated in accordance to Table 1. The maximum number of test function evaluations was used as a stopping criterion. Control and other parameter settings were as follows:

- Population size NP: 100
- External archive A of size H: $NP = 100$
- Dimension D: 10
- Runs R: 51
- Maximum number of test function evaluations $MAXTFE$: 10 000 $\times$ D = 100 000
- Number of generations G: $MAXTFE/NP$ = 1 000

5 Results

Statistical characteristics of the results are depicted in Tables 2, 3, 4, 5, 6 and 7, where the obtained values are differences from the global optimum of given function. The global optimum for each CEC2015 benchmark set function is equal to 100 $\times$ function

Table 2. Results of canonical SHADE algorithm with parent selection by PRNG with unform distribution on CEC2015 benchmark set functions. Best, Worst, Median, Mean and Std columns indicate lowest obtained value, highest obtained value, median, mean and standard deviation values of 51 independent runs.

f(x)	Best	Worst	Median	Mean	Std
$f(1)$	0.000	0.000	0.000	0.000	0.000
$f(2)$	0.000	0.000	0.000	0.000	0.000
$f(3)$	4.392	20.104	20.062	18.481	4.404
$f(4)$	0.119	4.389	3.065	2.784	0.809
$f(5)$	6.389	224.082	31.917	45.131	40.458
$f(6)$	0.000	120.165	0.680	6.039	23.311
$f(7)$	0.067	1.020	0.178	0.209	0.143
$f(8)$	0.020	1.542	0.478	0.473	0.339
$f(9)$	100.123	100.230	100.172	100.175	0.025
$f(10)$	216.537	219.001	216.537	216.640	0.430
$f(11)$	1.606	300.084	3.343	119.418	146.559
$f(12)$	100.908	101.830	101.460	101.445	0.209
$f(13)$	23.780	30.443	27.853	27.364	1.663
$f(14)$	2935.540	6996.830	2935.540	4267.120	1815.530
$f(15)$	100.000	100.000	100.000	100.000	0.000

Table 3. Results of SHADE algorithm with parent selection induced by Burgers CPRNG on CEC2015 benchmark set functions. Best, Worst, Median, Mean and Std columns indicate lowest obtained value, highest obtained value, median, mean and standard deviation values of 51 independent runs.

f(x)	Best	Worst	Median	Mean	Std
f(1)	0.000	0.000	0.000	0.000	0.000
f(2)	0.000	0.000	0.000	0.000	0.000
f(3)	4.294	20.108	20.065	18.856	3.779
f(4)	0.216	4.142	2.266	2.443	0.682
f(5)	9.270	273.329	28.102	48.405	52.241
f(6)	0.000	123.502	1.619	23.204	45.657
f(7)	0.057	1.130	0.180	0.333	0.349
f(8)	0.000	33.663	0.316	0.946	4.680
f(9)	100.119	100.251	100.167	100.170	0.033
f(10)	216.537	244.594	216.537	218.920	7.502
f(11)	1.663	300.099	3.507	131.079	148.657
f(12)	100.719	101.805	101.414	101.423	0.210
f(13)	24.072	32.040	28.093	27.810	1.593
f(14)	100.000	7181.850	2935.540	4004.460	2131.130
f(15)	100.000	100.000	100.000	100.000	0.000

Table 4. Results of SHADE algorithm with parent selection induced by Delayed Logistic CPRNG on CEC2015 benchmark set functions. Best, Worst, Median, Mean and Std columns indicate lowest obtained value, highest obtained value, median, mean and standard deviation values of 51 independent runs.

f(x)	Best	Worst	Median	Mean	Std
f(1)	0.000	0.000	0.000	0.000	0.000
f(2)	0.000	0.000	0.000	0.000	0.000
f(3)	4.643	20.100	20.052	18.015	4.529
f(4)	1.103	4.262	2.249	2.584	0.823
f(5)	7.054	158.830	32.663	47.632	38.754
f(6)	0.000	130.606	0.416	12.917	36.467
f(7)	0.072	1.020	0.180	0.228	0.208
f(8)	0.007	1.124	0.217	0.271	0.267
f(9)	100.114	100.204	100.167	100.165	0.024
f(10)	216.537	244.561	216.537	217.623	5.425
f(11)	2.050	300.084	3.971	136.997	149.281
f(12)	100.934	101.841	101.445	101.452	0.231
f(13)	24.472	32.518	28.168	28.362	1.760
f(14)	100.000	6997.340	2980.400	4602.500	2079.960
f(15)	100.000	100.000	100.000	100.000	0.000

Table 5. Results of SHADE algorithm with parent selection induced by Dissipative CPRNG on CEC2015 benchmark set functions. Best, Worst, Median, Mean and Std columns indicate lowest obtained value, highest obtained value, median, mean and standard deviation values of 51 independent runs.

f(x)	Best	Worst	Median	Mean	Std
f(1)	0.000	0.000	0.000	0.000	0.000
f(2)	0.000	0.000	0.000	0.000	0.000
f(3)	5.884	20.095	20.062	18.552	3.904
f(4)	1.055	4.399	2.432	2.623	0.846
f(5)	8.387	173.734	28.996	50.016	47.684
f(6)	0.012	120.160	1.153	8.400	28.118
f(7)	0.081	0.435	0.195	0.205	0.063
f(8)	0.027	1.178	0.399	0.399	0.266
f(9)	100.115	100.265	100.170	100.173	0.029
f(10)	216.537	216.537	216.537	216.537	0.000
f(11)	1.815	300.084	300.000	154.472	149.970
f(12)	100.806	101.931	101.405	101.398	0.234
f(13)	23.041	31.873	27.728	27.664	1.552
f(14)	2935.540	7036.980	2935.540	4267.530	1817.020
f(15)	100.000	100.000	100.000	100.000	0.000

Table 6. Results of SHADE algorithm with parent selection induced by Lozi CPRNG on CEC2015 benchmark set functions. Best, Worst, Median, Mean and Std columns indicate lowest obtained value, highest obtained value, median, mean and standard deviation values of 51 independent runs.

f(x)	Best	Worst	Median	Mean	Std
f(1)	0.000	0.000	0.000	0.000	0.000
f(2)	0.000	0.000	0.000	0.000	0.000
f(3)	3.409	20.101	20.061	18.709	4.210
f(4)	1.061	4.157	2.334	2.593	0.813
f(5)	10.607	171.623	36.862	55.358	45.664
f(6)	0.000	119.738	0.715	3.668	16.639
f(7)	0.085	0.444	0.199	0.212	0.075
f(8)	0.008	1.170	0.256	0.325	0.240
f(9)	100.110	100.230	100.173	100.176	0.028
f(10)	216.537	218.992	216.537	216.612	0.390
f(11)	1.391	300.084	3.576	125.277	147.710
f(12)	100.975	101.866	101.397	101.425	0.223
f(13)	23.238	35.003	27.874	27.681	1.828
f(14)	100.000	9945.420	2935.540	4587.500	2110.550
f(15)	100.000	100.000	100.000	100.000	0.000

Table 7. Results of SHADE algorithm with parent selection induced by Tinkerbell CPRNG on CEC2015 benchmark set functions. Best, Worst, Median, Mean and Std columns indicate lowest obtained value, highest obtained value, median, mean and standard deviation values of 51 independent runs.

f(x)	Best	Worst	Median	Mean	Std
f(1)	0.000	0.000	0.000	0.000	0.000
f(2)	0.000	0.000	0.000	0.000	0.000
f(3)	4.035	20.091	20.055	18.314	4.433
f(4)	0.422	4.247	3.058	2.764	0.809
f(5)	8.510	180.272	33.692	57.008	51.932
f(6)	0.000	120.941	0.416	5.309	23.500
f(7)	0.062	1.024	0.161	0.234	0.217
f(8)	0.000	17.245	0.316	0.635	2.386
f(9)	100.114	100.242	100.165	100.169	0.027
f(10)	216.537	245.331	216.537	217.939	5.523
f(11)	1.534	300.084	3.795	136.891	149.375
f(12)	100.918	101.736	101.416	101.385	0.219
f(13)	24.622	32.136	27.771	27.720	1.420
f(14)	100.000	7036.980	2935.540	4100.850	2326.540
f(15)	100.000	100.000	100.000	100.000	0.000

number (e.g. function 14 has the global optimum $f(x_0) = 1400$). Tables show lowest - best and highest - worst obtained values from 51 runs, median, mean and standard deviation values. Furthermore, each table represents the results of different CPRNG used for parent selection in the SHADE algorithm and also the results of canonical SHADE algorithm. The tables are ordered as follows – canonical SHADE, SHADE induced by Burgers, Delayed Logistic, Dissipative, Lozi and Tinkerbell CPRNG.

Additionally, the comparison of mean values of the canonical SHADE and SHADE induced by CPRNGs can be seen in Table 8, where best values are represented by bold numbers and if the canonical SHADE did not acquire the best mean result, Wilcoxon signed-rank test p-values between the canonical SHADE and CPRNG induced SHADE with the best mean result are displayed in the last column. The alternative hypothesis is that mean rank value of the canonical SHADE is greater than that of CPRNG induced SHADE.

Furthermore, the comparison between the canonical SHADE and SHADE induced by CPRNGs is illustrated in Figs. 1, 2, 3 and 4, where best value development over Test Function Evaluation (TFE) was averaged over all 51 runs. Figure 3 depicts the detail of the last 20 000 iterations of $f(9)$.

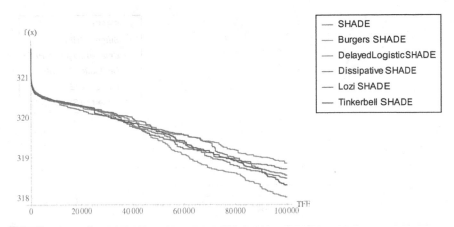

Fig. 1. Average best value development from 51 runs on $f(3)$, $D = 10$. (Color figure online)

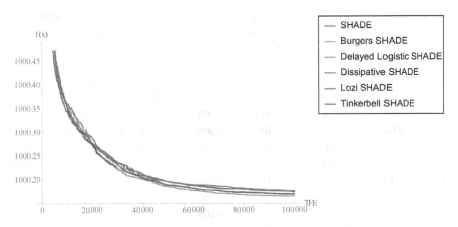

Fig. 2. Average best value development from 51 runs on $f(9)$, D = 10. (Color figure online)

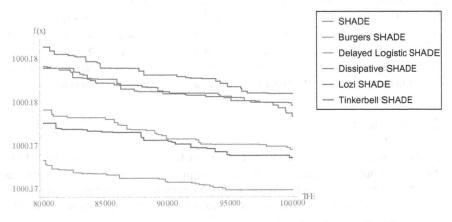

Fig. 3. Average best value development from 51 runs on $f(9)$, $D = 10$. Detail of the last 20 000 iterations. (Color figure online)

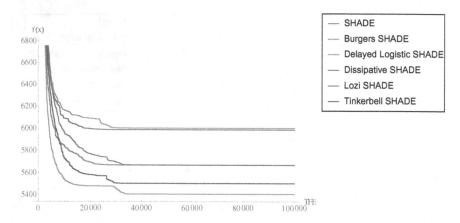

Fig. 4. Average best value development from 51 runs on *f*(14), *D* = 10. (Color figure online)

Table 8. Mean values of canonical SHADE and SHADE induced by five different CPRNGs on CEC2015 benchmark set functions, *p*-values of Wilcoxon signed-rank test between canonical SHADE and CPRNG induced SHADE with the best result on given function is given in the last column.

f(x)	SHADE	Burgers SHADE	DeLo SHADE	Dissipative SHADE	Lozi SHADE	Tinkerbell SHADE	p-value
f(1)	**0.000**	**0.000**	**0.000**	**0.000**	**0.000**	**0.000**	-
f(2)	**0.000**	**0.000**	**0.000**	**0.000**	**0.000**	**0.000**	-
f(3)	18.481	18.856	**18.015**	18.552	18.709	18.314	0.088
f(4)	2.784	**2.443**	2.584	2.623	2.593	2.764	0.010
f(5)	**45.131**	48.405	47.632	50.016	55.358	57.008	-
f(6)	6.039	23.204	12.917	8.400	**3.668**	5.309	0.298
f(7)	0.209	0.333	0.228	**0.205**	0.212	0.234	0.766
f(8)	0.473	0.946	**0.271**	0.399	0.325	0.635	0.000
f(9)	100.175	100.170	**100.165**	100.173	100.176	100.169	0.037
f(10)	216.640	218.920	217.623	**216.537**	216.612	217.939	0.037
f(11)	**119.418**	131.079	136.997	154.472	125.277	136.891	-
f(12)	101.445	101.423	101.452	101.398	101.425	**101.385**	0.065
f(13)	**27.364**	27.810	28.362	27.664	27.681	27.720	-
f(14)	4267.120	**4004.460**	4602.500	4267.530	4587.500	4100.850	0.350
f(15)	**100.000**	**100.000**	**100.000**	**100.000**	**100.000**	**100.000**	-

6   Result Discussion

As can be seen in Table 8, various CPRNGs used for parent selection in the SHADE algorithm can improve its performance on certain problems. This is true for functions *f* (3), *f*(4), *f*(8), *f*(9), *f*(10) and *f*(12) from CEC2015 benchmark set, where the obtained *p*-value from Wilcoxon signed-rank test indicates, that with the significance level set to

10 %, mean values reached by the SHADE algorithm with CPRNG are lower than that of the canonical SHADE algorithm. The performance on another set of functions $\{f(6), f(7), f(14)\}$ was improved as well, but the significance level was not reached. The canonical SHADE provided better mean value only on three of the fifteen test functions in benchmark set, $f(5)$, $f(11)$ and $f(13)$ and the same value was obtained on three test functions $f(1)$, $f(2)$ and $f(15)$.

Furthermore, the improvement in the speed of convergence can be seen in Figs. 2 and 4 which depict the average performance on functions $f(9)$ and $f(14)$ respectively. The same test function value can be reached in fewer evaluations which is crucial in real time optimization.

Additionally, Figs. 3 and 4 show that in the best value development on test functions $f(9)$ (detail of the last 20 000 iterations) and $f(14)$, significantly lower values were obtained by the SHADE algorithm with CPRNG parent selection. The results in Fig. 3 show, that three out of five chaotic systems used for parent selection are beneficial in terms of the reached test function value. This suggests, that not only one CPRNG, but their combination might be beneficial in the future multi-chaotic framework.

7 Conclusion

In this research paper, five different chaotic maps were used as a CPRNGs for parent selection process of the SHADE algorithm. The performance of the SHADE induced by CPRNG was tested on CEC2015 benchmark set and compared with the canonical SHADE.

In the previous research, CPRNGs were successfully used with a number of evolutionary algorithms - Self-Organizing Migrating Algorithm (SOMA) [15], canonical DE [13] and swarm intelligence algorithm - Particle Swarm Optimization (PSO) [12]. The aim of this research paper was to establish if inducing the SHADE algorithm parent selection phase by CPRNG will have any effect on the performance or if the effect will be neutralized by the adaptive behaviour of the SHADE algorithm. The overall performance was improved on the majority of the test functions but there is no universal winner between CPRNGs. Therefore, there is a space for the future research which would predict the most suitable CPRNG for given problem or possibly their combination and the product of this would be a multi-chaotic framework.

Thus the future research will be focused on the development, analysis and testing of such framework.

Acknowledgements. This work was supported by the Programme EEA and Norway Grants for funding via grant on Institutional cooperation project No. NF-CZ07-ICP-4-345-2016, also by Grant Agency of the Czech Republic – GACR P103/15/06700S, further by the Ministry of Education, Youth and Sports of the Czech Republic within the National Sustainability Programme Project No. LO1303 (MSMT-7778/2014). Also by the European Regional Development Fund under the Project CEBIA-Tech No. CZ.1.05/2.1.00/03.0089 and by Internal Grant Agency of Tomas Bata University under the Projects No. IGA/CebiaTech/2016/007.

References

1. Storn, R., Price, K.: Differential Evolution-A Simple and Efficient Adaptive Scheme for Global Optimization Over Continuous Spaces, vol. 3. ICSI, Berkeley (1995)
2. Price, K., Storn, R.M., Lampinen, J.A.: Differential Evolution: A Practical Approach To Global Optimization. Springer, Heidelberg (2006)
3. Cai, H.R., Chung, C.Y., Wong, K.P.: Application of differential evolution algorithm for transient stability constrained optimal power flow. IEEE Trans. Power Syst. **23**(2), 719–728 (2008)
4. Storn, R.: On the usage of differential evolution for function optimization. In: 1996 Biennial Conference of the North American Fuzzy Information Processing Society, NAFIPS, pp. 519–523. IEEE (1996)
5. Rogalsky, T., Kocabiyik, S., Derksen, R.W.: Differential evolution in aerodynamic optimization. Can. Aeronaut. Space J. **46**(4), 183–190 (2000)
6. Brest, J., Greiner, S., Bošković, B., Mernik, M., Zumer, V.: Self-adapting control parameters in differential evolution: a comparative study on numerical benchmark problems. IEEE Trans. Evol. Comput. **10**(6), 646–657 (2006)
7. Omran, M.G., Salman, A., Engelbrecht, A.P.: Self-adaptive differential evolution. In: Hao, Y., Liu, J., Wang, Y., Cheung, Y.M., Yin, H., Jiao, L., Ma, J., Jiao, Y.C. (eds.) CIS 2005, Part I. LNCS, vol. 3801, pp. 192–199. Springer, Heidelberg (2005)
8. Qin, A.K., Huang, V.L., Suganthan, P.N.: Differential evolution algorithm with strategy adaptation for global numerical optimization. IEEE Trans. Evol. Comput. **13**(2), 398–417 (2009)
9. Zhang, J., Sanderson, A.C.: JADE: adaptive differential evolution with optional external archive. IEEE Trans. Evol. Comput. **13**(5), 945–958 (2009)
10. Tanabe, R., Fukunaga, A.: Success-history based parameter adaptation for differential evolution. In: 2013 IEEE Congress on Evolutionary Computation (CEC), pp. 71–78. IEEE (2013)
11. Skanderova, L., Zelinka, I., Šaloun, P.: Chaos powered selected evolutionary algorithms. In: Zelinka, I., Chen, G., Rössler, O.E., Snasel, V., Abraham, A. (eds.) Nostradamus 2013: Prediction, Modeling and Analysis of Complex Systems. Advances in Intelligent Systems and Computing, vol. 210, pp. 111–124. Springer, Switzerland (2013)
12. Pluhacek, M., Senkerik, R., Zelinka, I.: Particle swarm optimization algorithm driven by multichaotic number generator. Soft. Comput. **18**(4), 631–639 (2014)
13. Senkerik, R., Pluhacek, M., Kominkova Oplatkova, Z., Davendra, D.: On the parameter settings for the chaotic dynamics embedded differential evolution. In: 2015 IEEE Congress on Evolutionary Computation (CEC), pp. 1410–1417. IEEE (2015)
14. Caponetto, R., Fortuna, L., Fazzino, S., Xibilia, M.G.: Chaotic sequences to improve the performance of evolutionary algorithms. IEEE Trans. Evol. Comput. **7**(3), 289–304 (2003)
15. Davendra, D., Zelinka, I., Senkerik, R.: Chaos driven evolutionary algorithms for the task of PID control. Comput. Math. Appl. **60**(4), 1088–1104 (2010)
16. Chen, Q., Liu, B., Zhang, Q., Liang, J.J., Suganthan, P.N., Qu, B.Y.: Problem definition and evaluation criteria for CEC 2015 special session and competition on bound constrained single-objective computationally expensive numerical optimization. Technical report, Computational Intelligence Laboratory, Zhengzhou University, China and Nanyang Technological University, Singapore (2014)
17. Sprott, J.C., Sprott, J.C.: Chaos and Time-Series Analysis, vol. 69. Oxford University Press, Oxford (2003)

Tabu Search Hybridized
with Multiple Neighborhood Structures
for the Frequency Assignment Problem

Khaled Alrajhi$^{(\boxtimes)}$, Jonathan Thompson, and Wasin Padungwech

School of Mathematics, Cardiff University, Cardiff, UK
{alrajhika,thompsonjm1,padungwechw}@cardiff.ac.uk

Abstract. This study proposes a tabu search hybridized with multiple neighborhood structures to solve a variant of the frequency assignment problem known as the minimum order frequency assignment problem. This problem involves assigning frequencies to a set of requests while minimizing the number of frequencies used. Several novel and existing techniques are used to improve the efficiency of this algorithm. This includes a novel technique that aims to determine a lower bound on the number of frequencies required from each domain for a feasible solution to exist, based on the underlying graph coloring model. These lower bounds ensure that the search focuses on parts of the solution space that are likely to contain feasible solutions. Our tabu search algorithm was tested on real and randomly generated benchmark datasets of the static problem and achieved competitive results.

1 Introduction

The frequency assignment problem (FAP) is related to wireless communication networks, which are used in many applications such as mobile phones, TV broadcasting and Wi-Fi. The aim of the FAP is to assign frequencies to wireless communication connections (also known as requests) while satisfying a set of constraints, which are usually related to prevention of a loss of signal quality. Note that the FAP is not a single problem. Rather, there are variants of the FAP that are encountered in practice. The minimum order FAP (MO-FAP) is the first variant of the FAP that was discussed in the literature, and was brought to the attention of researchers by [1]. In the MO-FAP, the aim is to assign frequencies to requests in such a way that no interference occurs, and the number of used frequencies is minimized. As the MO-FAP is NP-complete [2], it is usually solved by meta-heuristics.

Many meta-heuristics have been proposed to solve the MO-FAP including genetic algorithm (GA) [3], evolutionary search (ES) [4], ant colony optimization (ACO) [5], simulated annealing (SA) [6] and tabu search (TS) [6–9]. It can be seen from the literature that TS is a popular meta-heuristic for solving difficult combinatorial optimization problems. This generally applicable algorithm has proved to be an efficient way of finding a high quality solution for a variety of

© Springer International Publishing Switzerland 2016
M.J. Blesa et al. (Eds.): HM 2016, LNCS 9668, pp. 157–170, 2016.
DOI: 10.1007/978-3-319-39636-1_12

optimization problems e.g. [10]. However, existing algorithms in the literature are unable to find optimal solutions in some datasets for the MO-FAP.

In this paper, we present an improved TS algorithm hybridized with multiple neighborhood structures, one of which is used as a diversification technique. The concept of using multiple neighborhood structures is inherited from the variable neighborhood search algorithm, introduced by [11]. In contrast, other TS algorithms for the MO-FAP in the literature implement only a single neighborhood structure, e.g. [6–9]. Another new technique used in our TS algorithm is applying a lower bound on the number of frequencies that are required from each domain for a feasible solution to exist, based on the underlying graph coloring model. This ensures that the search focuses on parts of the solution space that are likely to contain feasible solutions. In this study, experiments were carried out on CELAR and GRAPH datasets[1], and the results show that our TS algorithm outperforms other algorithms in the literature.

This paper is organized as follows: the next section gives an overview of the MO-FAP. Section 3 explains how the underlying graph coloring model for the MO-FAP can be used to provide a lower bound on the number of frequencies and how this information can then be used to assist the search. In Sects. 4 and 5, the description of the TS algorithm for the MO-FAP is given. In Sect. 6, the results of this algorithm are given and compared with those of other algorithms in the literature, before this study finishes with conclusions and future work.

2   Overview of the MO-FAP

The main concept of the MO-FAP is assigning a frequency to each request while satisfying a set of constraints and minimizing the number of used frequencies. The MO-FAP can be defined formally as follows: Given a set of requests $R = \{r_1, r_2, \ldots, r_{NR}\}$ and a set of frequencies $F = \{f_1, f_2, \ldots, f_{NF}\} \subset \mathbb{Z}^+$, where NR is the number of requests and NF is the number of frequencies, and a set of constraints related to the requests and frequencies, the goal is to assign one frequency to each request so that the given set of constraints are satisfied and the number of frequencies used is minimized. The frequency that is assigned to request r_i is denoted by f_{r_i} throughout of this study. The MO-FAP has four types of constraints:

1. *Bidirectional Constraints*: this type of constraint forms a link between each pair of requests $\{r_{2i-1}, r_{2i}\}$, where $i = 1, \ldots, NR/2$. In these constraints, the frequencies $f_{r_{2i-1}}$ and $f_{r_{2i}}$ should be distance $d_{r_{2i-1}r_{2i}} \in \mathbb{Z}^+$ apart. In the datasets considered here, $d_{r_{2i-1}r_{2i}}$ is always equal to a constant value (238). These constraints can be written as follows:

$$|f_{r_{2i-1}} - f_{r_{2i}}| = d_{r_{2i-1}r_{2i}} \quad \text{for } i = 1, \ldots, NR/2 . \tag{1}$$

---

[1] These are available at http://fap.zib.de/problems/CALMA/ (last accessed 25 December 2015).

2. *Interference Constraints*: this type of constraint forms a link between a pair of requests $\{r_i, r_j\}$. The frequencies f_{r_i} and f_{r_j} should be more than distance $d_{r_i r_j} \in \mathbb{Z}^+$ apart. These constraints can be written as follows:

$$|f_{r_i} - f_{r_j}| > d_{r_i r_j} \quad \text{for } 1 \leq i < j \leq NR \ . \tag{2}$$

3. *Domain Constraints*: the set of available frequencies for each request r_i is denoted by the domain $D_{r_i} \subset F$, where $\cup_{r_i \in R} D_{r_i} = F$. Hence, the frequency which is assigned to r_i must belong to D_{r_i}. For the datasets considered in this study, there are 7 available domains.
4. *Pre-assignment Constraints*: for certain requests, the frequencies have already been pre-assigned to given values.

3 Graph Coloring Model for the MO-FA

The graph coloring problem (GCP) is an underlying model to the MO-FAP [12]. The GCP involves allocating a color to each vertex such that no adjacent vertices are in the same color class and the number of colors is minimized. The MO-FAP can be represented as a GCP by representing each request as a vertex and a bidirectional or an interference constraint as an edge joining the corresponding vertices.

One useful concept of graph theory is the idea of cliques. A clique can be defined as a set of vertices in which each vertex is linked to all other vertices. A maximum clique is the largest among all cliques in a graph. Vertices in a clique have to be allocated to different colors in a feasible coloring. Therefore, the size of the maximum clique acts as a lower bound on the minimum number of colors and therefore, by extension, as a lower bound on the number of frequencies for the MO-FAP. For example, the requests $r_1, r_{200}, r_{871}, r_{872}$ and r_{899} form a clique in the CELAR 01 instance (see Fig. 1). All of these requests are linked to each other by either a bidirectional or an interference constraint.

Figure 1 shows 5 different requests forming a clique, so at least 5 different frequencies are required. As the requests belong to different domains, the graph coloring model for each domain can be considered separately and then a lower bound on the number of frequencies that is required from each domain can also be calculated.

A Branch and Bound algorithm is used to obtain the size of a maximum clique. Table 1 gives a lower bound of the number of frequencies required from each domain and whole instance, and the time taken to calculate the lower bounds.

4 Overview of the Tabu Search Algorithm

A key decision when designing TS is the definition of the solution space and the corresponding cost function.

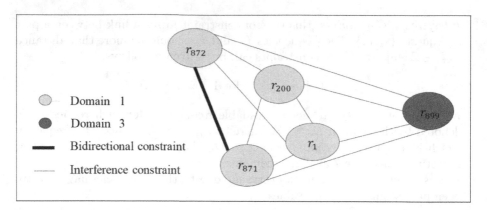

Fig. 1. An example of a clique in the CELAR 01 instance in the graph coloring model.

Table 1. Lower bounds of the numbers of frequencies required from each domain and whole instance and time taken to calculate the lower bounds.

Instance	Domain							Whole instance	Run time
	1	2	3	4	5	6	7		
CELAR 01	10	9	10	4	4	7	2	12	1.50 sec
CELAR 02	10	0	10	0	0	0	2	14	0.02 sec
CELAR 03	10	0	10	0	2	0	2	12	0.06 sec
CELAR 04	10	0	10	4	2	0	2	44	0.34 sec
CELAR 11	20	0	14	4	2	0	2	20	0.34 sec
GRAPH 01	8	3	6	2	4	4	2	18	0.03 sec
GRAPH 02	6	2	4	0	2	4	0	14	0.12 sec
GRAPH 08	10	2	6	2	3	8	3	16	0.28 sec
GRAPH 09	6	2	10	2	2	8	2	18	0.48 sec
GRAPH 14	6	2	4	2	0	2	2	8	0.48 sec

4.1   Solution Space and Cost Function

In most cases, it is relatively straightforward to find solutions that satisfy the bidirectional, the domain and the pre-assignment constraints, and to define a neighborhood operator that moves between such solutions [13]. Hence, the solution space here is defined as the set of all possible assignments satisfying all of those constraints. Note that the interference constraints are relaxed because these are the most difficult to be satisfied. The cost function is defined as the number of broken interference constraints, also known as the number of violations. One of the advantages of this configuration is that the number of requests is halved because each request is linked with another request based on the bidirectional constraints. This configuration was used in [6,8].

A further approach is considered where the solution space consists of solutions that satisfy only the domain and the pre-assignment constraints, while the cost function counts the number of broken bidirectional and interference constraints. This configuration was used in [7,14].

The solution space could have been defined as the set of all possible feasible assignments, that is, satisfying all of the constraints, and the corresponding cost function as the number of used frequencies. However, there are a number of difficulties with this configuration: TS has been found to be poor with this configuration [7]. Moreover, it may be difficult to move from one feasible solution to another. Furthermore, a large number of neighbor solutions with the same cost may differ greatly in their quality [15]. Hence, this is not considered in this study.

4.2   Sub-problem in the MO-FAP

Using the solution space which relaxes some constraints creates the following sub-problem: minimizing the number of violations with a fixed number of used frequencies. If a solution with zero violations, i.e. a feasible solution, is found, then the number of used frequencies is reduced in the creating violations phase (see Sect. 5.5) and the sub-problem is reconsidered. The process is repeated until a feasible solution can no longer be found. This process is similar to [16] in their TS algorithm for the GCP and [6,8] in their TS algorithm for the MO-FAP.

4.3   Structure of the Tabu Search Algorithm

Our TS algorithm consists of three phases, namely the initial solution phase, the creating violations phase and the improvement phase. The initial solution phase (see Sect. 5.4) generates an initial solution. Assume the initial solution is feasible. Then, the creating violations phase (see Sect. 5.5) reduces the number of used frequencies by removing a pair of used frequencies. Then, all pairs of requests that are assigned to the removed pair of frequencies are re-assigned to another pair of used frequencies, which may result in some violations. The improvement phase (see Sect. 5.6) aims to reduce the number of violations to zero, using three neighborhood structures. If this results in a feasible solution within a specified number of iterations, then the creating violation phase is revisited to remove another pair of used frequencies. After that, the process continues until either no feasible solution can be found, at which time the process is terminated and the feasible solution in the previous iteration is returned, or the optimal solution is found. In case the initial solution is not feasible, the creating violations phase can be omitted and the search moves immediately to the improvement phase.

The overall structure of our TS algorithm for the MO-FAP can be described as follows.

1. Implement the initial solution phase.
2. If the number of violations equals 0, go to step 3. Else, go to step 5.
3. If the number of used frequencies equals the lower bound, return the current feasible solution. Else, go to step 4.

4. Implement the creating violations phase.
5. Implement the improvement phase.
6. If the number of violations equals 0, go to step 3. Else, go to step 7.
7. If the previous solution is feasible, return it. Else, return the current infeasible solution.

5   Components of the Tabu Search Algorithm

Throughout this section, all the constraints except the interference constraints are regarded as hard constraints. That is, we use the first configuration described in Sect. 4.1.

5.1   Neighborhood Structures

Three different neighborhood structures are considered:

- *Move Neighborhood Structure (MNS)*: this structure is defined as the set of solutions obtained by selecting a pair of requests and re-assigning them to a different pair of used frequencies while satisfying all the hard constraints. This neighborhood investigates all the possible moves for all pairs of requests and used frequencies (the maximum possible number of such moves is $NR \times n_f$, where n_f is the number of used frequencies). This ensures that the number of used frequencies does not increase. This structure is simple and commonly used for TS algorithms in the literature e.g. [6–8].
- *Swap Neighborhood Structure (SNS)*: this structure is defined as the set of solutions obtained by swapping the frequencies of a pair of requests. SNS proves to be quick as it contains a small number of neighbors (at most $NR/2$), yet it can improve the solution quality.
- *Diversification Neighborhood Structure (DNS)*: this structure, unlike the previous structures, is intended to diversify the search, i.e. move to a different part of the solution space. It consists of the set of solutions obtained by replacing a pair of used (old) frequencies with a pair of unused (new) frequencies. Given a pair of old frequencies, another pair of frequencies is accepted if it can be assigned to all pairs of requests which were assigned to the old pair without breaking any hard constants. However, any re-assignment that causes the number of used frequencies to drop below the lower bound for some domains (see Sect. 3) is not considered.

5.2   Tabu Lists

In our TS algorithm, three independent tabu lists, one for each neighborhood structure, are defined. Notice that all of the tabu lists are cleared after the sub-problem is solved. These tabu lists are described as follows:

- *Move Tabu List*: when a pair of requests is assigned to another pair of frequencies, then the pair of requests and the removed pair of frequencies are added to the tabu list and this assignment is classified as forbidden for a given number of iterations (i.e. tabu tenure).
- *Swap Tabu List*: when a pair of requests are swapped, then this pair is added to the swap tabu list. This list prevents a pair of requests from being swapped more than once.
- *Diversification Tabu List*: when a pair of old frequencies is replaced by a pair of new frequencies, then both of them are added to the diversification tabu list.

5.3   Aspiration Criteria

The tabu lists can be too restrictive by forbidding some attractive moves even when there is no harm of cycling. Therefore, it can be beneficial to ignore the tabu lists. This is called an aspiration criterion. Here, the most commonly used aspiration criterion is applied, that is, to accept a tabu move if it leads to a better solution than the current best found one.

5.4   The Initial Solution Phase

An initial solution is generated by the following greedy algorithm: a pair of requests which has the smallest number of feasible pairs of frequencies is selected. Then, among those pairs of frequencies, the one which is feasible for most pairs of requests is assigned to the selected pair of requests. If there are no feasible pairs of frequencies, a pair of frequencies is randomly selected. In case the initial solution is infeasible, a descent method with MNS (see Sect. 5.1) is used to reduce the number of violations.

5.5   The Creating Violations Phase

This phase aims to reduce the number of used frequencies in a feasible solution by removing a pair of frequencies. The removed pair of frequencies must satisfy two conditions: First, neither of the frequencies are required to satisfy any pre-assignment constraints. Second, the lower bound for each domain (see Sect. 3) must be satisfied after deleting these frequencies. If there is more than one candidate pair of frequencies, then the one which is assigned to the least number of pairs of requests is selected. If there is still more than one such pair, then one of them is selected randomly. After that, the pairs of requests assigned to the candidate pair of frequencies are re-allocated to a feasible pair of used frequencies. If there is no feasible pair of used frequencies, then these requests are re-allocated to an infeasible pair of used frequencies at random. In case this process leads to a feasible solution, then a further pair is removed. Otherwise, the improvement phase (see Sect. 5.6) is executed to find a feasible solution. The creating violations phase was previously used in [8].

5.6   The Improvement Phase

The aim of this phase is to solve the sub-problem using the three neighborhood structures.

Ordering of Neighborhood Structures. The iterative procedure of our TS algorithm starts in the improvement phase. The improvement phase consists of three neighborhood structures (MNS, SNS and DNS). In MNS and SNS, only used frequencies are considered, while DNS considers only unused frequencies. MNS is explored first because it contains a large number of neighbors. SNS, which covers a limited number of neighbors, is then considered to support the MNS. DNS aims to jump from the current position in the solution space to a new position by removing a pair of used frequencies and adding a new one from the set of pair of unused frequencies. Therefore, DNS is intended to diversify the search rather than reduce the number of violations, which reflects the reason for leaving it as the last structure.

Implementation of the Improvement Phase. Each iteration involves one of the three neighborhood structures. This phase begins with MNS. If this structure results in a better solution, then it is accepted. Otherwise, it is repeated until the structure is executed for a given number of times consecutively without improvement. Then, the search enters SNS. If this structure leads to a better or equally good solution, then the search goes back to MNS. Otherwise, it appears there is little prospect of finding a better solution in the current region of the solution space, so the search enters DNS. A solution from DNS is accepted and the search returns to MNS.

It was found that on occasions, no moves in DNS are allowed due to the tabu lists, the pre-assignment constraints and the lower bound for each domain. If this happens, the criteria of selecting a pair of new frequencies in DNS will be modified. A pair of frequencies is accepted as a pair of new frequencies if it can be allocated to at least one pair (instead of all pairs) of requests assigned to the pair of old frequencies. Although the pair of new frequencies will not be allowed to be removed because of the diversification tabu list, the pair of old frequencies will be allowed to return to the solution because of a limited number of neighbors in this structure.

The output of the improvement phase can be a feasible or an infeasible solution. If it is a feasible, but not optimal solution, then the algorithm returns to the creating violations phase. On the other hand, if the output is an infeasible solution, then the algorithm returns to MNS. This continues until one of the stopping criteria is satisfied.

5.7   Stopping Criteria

Our TS algorithm has three different stopping criteria as follows: (i) the feasible solution whose number of frequencies is equal to the lower bound is found (as this is the optimal solution), (ii) the number of iterations is equal to a given number, (iii) the DNS is executed for a certain number of times.

6   Experiments and Results

This section firstly provides the results of our TS algorithm for the MO-FAP using CELAR and GRAPH datasets (available on the FAP website[2]). Secondly, the process of our TS algorithm is discussed and analyzed. Finally, the performance of our TS algorithm is compared with other algorithms in the literature.

Table 2 presents details of the MO-FAP datasets considered in this study including the numbers of requests and constraints for each instance.

Table 2. Details of CELAR and GRAPH datasets.

Instance	No. of requests	No. of bidirectional constraints	No. of interference constraints	No. of domain constraints	No. of pre-assignment constraints	Total no. of constraints
CELAR 01	916	458	5,090	916	0	6,464
CELAR 02	200	100	1,135	200	0	1,435
CELAR 03	400	200	2,560	400	0	3,160
CELAR 04	680	340	3,627	400	280	4,647
CELAR 11	680	340	3,763	680	0	4,783
GRAPH 01	200	100	1,034	200	0	1,334
GRAPH 02	400	200	2,045	400	0	2,645
GRAPH 08	680	340	3,417	680	0	4,437
GRAPH 09	916	458	4,788	916	0	6,162
GRAPH 14	916	458	4,180	916	0	5,554

Based on experimentations, the parameters of our TS algorithm are set as follows:

- The maximum number of iterations is 10,000.
- The maximum number of times of accepting worse solutions consecutively in MNS is 100.
- The maximum number of times of executing DNS is 20.
- The tabu tenure of the move tabu list is 100.
- The tabu tenure of the swap tabu list is $NR/2$.
- The tabu tenure of the diversification tabu list is 20.

In this study, the algorithm was coded using FORTRAN 95 and all experiments were conducted on a 3.0 GHz Intel Core I3-2120 Processor (2nd Generation) with 8GB RAM and a 1TB Hard Drive.

---

[2] http://fap.zib.de/problems/CALMA/ (last accessed 25 December 2015).

6.1 Results Comparison of the Tabu Search Algorithm

This section provides the results of our TS algorithm for the MO-FAP. Five runs are performed for each instance, and each run uses a different random number stream. The results include the number of used frequencies in the best, the worst and the average solutions (with the optimal ones shown in bold), the average run time for each instance and the optimal solutions (known and available on the FAP website[3]). Note that the run time of finding the lower bound of the number of frequencies for each domain (given in Table 1) is included.

Table 3 shows that TS achieved optimal solution for all the instances except CELAR 11 and the solutions were obtained in a reasonable time, mostly less than 5 min.

Table 3. Results of the TS algorithm for the MO-FAP.

Instance	Best found	Worst found	Average solution	Average time	Worst time	Optimal solution
CELAR 01	**16**	**16**	**16**	3.63 min	4.01 min	16
CELAR 02	**14**	**14**	**14**	0.52 sec	0.61 sec	14
CELAR 03	**14**	16	14.8	1.00 min	1.20 min	14
CELAR 04	**46**	**46**	**46**	54.34 sec	1.18 min	46
CELAR 11	38	40	38.4	8.81 min	9.11 min	22
GRAPH 01	**18**	**18**	**18**	5.43 sec	7.09 sec	18
GRAPH 02	**14**	**14**	**14**	2.16 sec	4.98 sec	14
GRAPH 08	**18**	**18**	**18**	24.28 sec	33.12 sec	18
GRAPH 09	**18**	**18**	**18**	3.01 min	4.91 min	18
GRAPH 14	**8**	**8**	**8**	4.81 min	5.02 min	8

A further approach is considered where the bidirectional constraints are not enforced and the solution space consists of solutions that satisfy only the domain and the pre-assignment constraints, and the cost function counts the number of broken bidirectional and interference constraints. Experiments show that this approach did not lead to good results compared with the former one. This shows that enforcing bidirectional constraints is an important factor in improving the search efficiency.

6.2 Analysis of Implementation Process

In this section, the process of our TS algorithm is discussed and analyzed. Figure 2 shows the number of used frequencies and the number of violations throughout one run using the CELAR 01 instance.

---

[3] http://fap.zib.de/problems/CALMA/ (last accessed 25 December 2015).

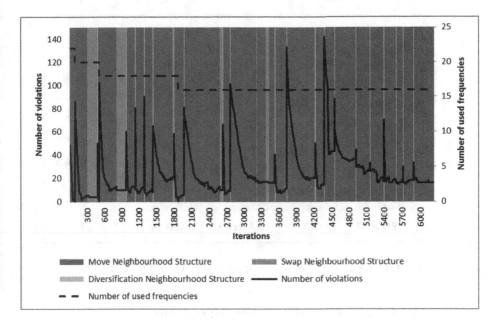

Fig. 2. Numbers of used frequencies and violations throughout one run on CELAR 01.

Figure 2 shows that TS start with an initial feasible solution using 22 frequencies and this number was reduced to 16 frequencies. Although all neighborhood structures have been involved during the process of this algorithm, the most executed structure is MNS, which is represented by the red color. This justifies the fact that this structure is the most successful and commonly used for TS in the literature. SNS came as the second most executed structure. This reflects the limitation of this structure and its objective, which is to support MNS. DNS is executed in a limited number of times and most of the times it results in an increase in the number of violations. This agrees with the aim of this structure, which is to diversify the search rather than optimize it.

In order to investigate the importance of each neighborhood structure, four different approaches of our TS algorithm are compared: Approach 1 applies the initial solution phase only; Approach 2 applies MNS only; Approach 3 applies MNS and SNS only and Approach 4 applies all the neighborhood structures.

Figure 3 presents the results of four different approaches to a selection instances (specifically, CELAR 01, CELAR 03, GRAPH 09 and GRAPH 14). The selected instances are chosen to represent different numbers of requests and constraints.

It can be seen from Fig. 3 that all the neighborhood structures play a role to achieve the goal. All the instances achieve a better solution after adding each neighborhood structure, i.e. all the neighborhood structures are essential to solve the problem.

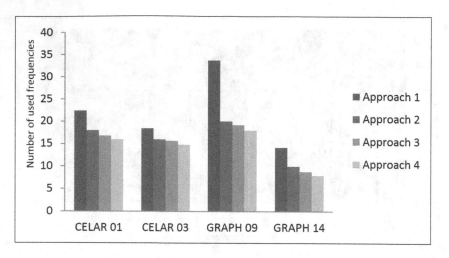

Fig. 3. Comparison of different approaches of our TS algorithm.

6.3 Results Comparison with Other Algorithms

In this section, the results of our TS algorithm and other algorithms in the literature are compared. Table 4 shows the best found results, where the result shown in bold means these reach the optimal solution and a dash "-" means that the result is not available.

Table 4. Results of our TS algorithm and other algorithms in the literature.

Instance	GA [3]	ES [4]	SA [6]	TS [6]	TS [9]	Our TS	Optimal solution
CELAR 01	20	-	**16**	**16**	18	**16**	16
CELAR 02	**14**	**14**	**14**	**14**	**14**	**14**	14
CELAR 03	16	**14**	**14**	**14**	**14**	**14**	14
CELAR 04	**46**	-	**46**	**46**	**46**	**46**	46
CELAR 11	32	-	24	**22**	24	38	22
GRAPH 01	20	**18**	-	**18**	**18**	**18**	18
GRAPH 02	16	**14**	-	**14**	16	**14**	14
GRAPH 08	-	-	-	20	24	**18**	18
GRAPH 09	28	-	-	22	22	**18**	18
GRAPH 14	14	-	-	10	12	**8**	8

It can be seen from Table 4 that our TS algorithm achieved the best results compared with those of other algorithms in the literature. In fact, it achieved the optimal solution for all the instances except for CELAR 11. Moreover, it is the only algorithm in Table 4 that achieved the optimal solution for GRAPH

08, GRAPH 09 and GRAPH 14. However, a better result for CELAR 11 (the optimal solution) was found using the TS algorithm in [6]. Overall, our TS algorithm showed competitive results compared with those of other algorithms in the literature.

7   Conclusions and Future Work

In this paper, we have presented an improved TS algorithm for solving the MO-FAP. Several techniques have been used to improve the performance of this algorithm. These include hybridizing TS with three different neighborhood structures, one of which is used as a diversification technique, and using a lower bound for each domain based on the underlying graph coloring model. Moreover, based on the definition of the solution space which relaxes some constraints, a sub-problem of minimizing the number of violations is considered to find a feasible solution with a fixed number of used frequencies after the creating violations phase. Based on the results comparison, our TS algorithm outperformed other algorithms in the literature.

Clearly, there are many other variants of TS that could have been assessed. For example, a more advanced neighborhood structure could be used such as swapping pairs of requests with each other or forming chains similar to Kempe Chains in the GCP. Further investigations of these are left as future work.

References

1. Metzger, B.H.: Spectrum management technique presented at 38th national orsa meeting. Detroit, MI (Fall 1970) (1970)
2. Garey, M.R., Johnson, D.S.: A Guide to the Theory of NP-Completeness. WH Freemann, New York (1979)
3. Kapsalis, A., Chardaire, P., Rayward-Smith, V.J., Smith, G.D.: The radio link frequency assignment problem: a case study using genetic algorithms. In: Fogarty, T.C. (ed.) AISB-WS 1995. LNCS, vol. 993, pp. 117–131. Springer, Heidelberg (1995)
4. Crisan, C., Mühlenbein, H.: The frequency assignment problem: a look at the performance of evolutionary search. In: Hao, J.-K., Lutton, E., Ronald, E., Schoenauer, M., Snyers, D. (eds.) AE 1997. LNCS, vol. 1363, pp. 263–273. Springer, Heidelberg (1998)
5. Parsapoor, M., Bilstrup, U.: Ant colony optimization for channel assignment problem in a clustered mobile ad hoc network. In: Tan, Y., Shi, Y., Mo, H. (eds.) ICSI 2013, Part I. LNCS, vol. 7928, pp. 314–322. Springer, Heidelberg (2013)
6. Tiourine, S.R., Hurkens, C.A.J., Lenstra, J.K.: Local search algorithms for the radio link frequency assignment problem. Telecommun. Syst. **13**(2–4), 293–314 (2000)
7. Bouju, A., Boyce, J.F., Dimitropoulos, C.H.D., Vom Scheidt, G., Taylor, J.G.: Tabu search for the radio links frequency assignment problem. Applied Decision Technologies (ADT-95) (London) (1995)
8. Hao, J.-K., Dorne, R., Galinier, P.: Tabu search for frequency assignment in mobile radio networks. J. Heuristics **4**(1), 47–62 (1998)

9. Bouju, A., Boyce, J.F., Dimitropoulos, C.H.D., Vom Scheidt, G., Taylor, J.G., Likas, A., Papageorgiou, G., Stafylopatis, A.: Intelligent search for the radio link frequency assignment problem. In: Proceedings of the International Conference on Digital Signal Processing, Cyprus (1995)
10. Glover, F., Laguna, M.: Tabu search applications. In: Glover, F., Laguna, M. (eds.) Tabu Search, pp. 267–303. Springer, Heidelberg (1997)
11. Mladenović, N., Hansen, P.: Variable neighborhood search. Comput. Oper. Res. **24**(11), 1097–1100 (1997)
12. Hale, W.K.: Frequency assignment: theory and applications. Proc. IEEE **68**(12), 1497–1514 (1980)
13. Dorne, R., Hao, J.-K.: Constraint handling in evolutionary search: a case study of the frequency assignment. In: Ebeling, W., Rechenberg, I., Voigt, H.-M., Schwefel, H.-P. (eds.) PPSN 1996. LNCS, vol. 1141, pp. 801–810. Springer, Heidelberg (1996)
14. Hao, J.-K., Perrier, L.: Tabu search for the frequency assignment problem in cellular radio networks. Technical report LGI2P, EMA-EERIE, Parc Scientifique Georges Besse, Nimes, France (1999)
15. Dowsland, K.A., Thompson, J.M.: An improved ant colony optimisation heuristic for graph colouring. Discrete Appl. Math. **156**(3), 313–324 (2008)
16. Hertz, A., de Werra, D.: Using tabu search techniques for graph coloring. Computing **39**(4), 345–351 (1987)

The Capacitated m Two-Node Survivable Star Problem: A Hybrid Metaheuristic Approach

Gabriel Bayá$^{(\boxtimes)}$, Antonio Mauttone, Franco Robledo, and Pablo Romero

Departamentp de Investigación Operativa, Universidad de la República,
Montevideo, Uruguay
{gbaya,mauttone,frobledo,promero}@fing.edu.uy

Abstract. In telecommunications, a traditional method to connect multiterminal systems is to use rings. The goal of the Capacitated m Ring Star Problem (CmRSP) is to connect terminals by m rings which meet at a distinguished node, and possibly by some pendant links, at minimum cost. In this paper, we introduce a relaxation for the CmRSP, called Capacitated m Two-Node Survivable Star Problem (CmTNSSP for short). The CmTNSSP belongs to the $\mathcal{NP}$-Hard class of computational problems. Therefore, we address a GRASP hybrid metaheuristic which alternates local searches that obtain incrementally better solutions, and exact resolution local searches based on Integer Linear Programming models. In consonance with predictions provided by Clyde Monma, the network can be equally robust but cheaper than in the original CmRSP.

Keywords: Network optimization · CmRSP · CmTSSP · Hybrid metaheuristics · GRASP · VND · ILP

1 Motivation

A natural approach to reach two-node connectivity is to connect all terminals in a ring or cycle in an economic way. In this scenario nodes are connected to one another by two independent paths. This problem is called Traveling Salesman Problem, and it is widely studied in the scientific literature. Clyde Monma et al. [9] described what is considered to be a cornerstone in the area of topological network design. They proved that a minimum-cost 2-node-connected metric network is either a Hamiltonian Tour or presents a special graph topology as an induced subgraph. This topology is sketched in Fig. 1; it was refered to as Monma graphs for the first time in [4]. We will stick to this terminology. In the physical design of a telephony deployment, it is useful to consider several 2-node-connected components joined to a perfect telephone exchange, and to connect some distant terminal nodes to some ring. A cost-effective "shape" of a solution is provided by Roberto Baldacci et al. [1]. We are given a distinguished node (or *depot*), several terminal nodes and optional nodes. In order to connect all terminals, the authors propose to find the cheapest structure of m rings which share the depot, while some terminals can be pendant on some node of a ring.

© Springer International Publishing Switzerland 2016
M.J. Blesa et al. (Eds.): HM 2016, LNCS 9668, pp. 171–186, 2016.
DOI: 10.1007/978-3-319-39636-1_13

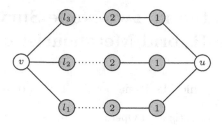

Fig. 1. Monma's graph structure.

The number of nodes within a ring must not exceed the depot capacity, and the cost of pendant edges is different than the cost of the edges within the rings. The minimum-cost design of the structure composed by the m rings and pendant nodes is called Capacitated m Ring Star Problem, termed herein CmRSP for short.

Inspired by the potential savings predicted by Clyde Monma et al., and supported by their theorem where the cost of the best ring could be even 4/3 times larger than the cost of the best 2-node-connected topology, we relaxed the condition of rings and considered arbitrary 2-node-connected components instead. The goal of this paper is to design a resilient cost-effective network from a topological stand point, suitable for delay sensitive applications on an Internet infrastructure. The main contributions are the following:

- The Capacitated m Two-Node Survivable Star Problem (CmTNSSP) is introduced.
- Given its intractability, a heuristic resolution is developed. We adopted a GRASP approach enriched with a Variable Neighborhood Descent, or GRASP-VND using some local searches based on Integer Linear Programming.
- A fair comparison with prior works in the area promotes the design of arbitrary 2-node-connected components, instead of rings (which were previously used by Baldacci et al.).

This article is organized in the following manner. The formal definitions for both problems, namely CmRSP and CmTNSSP, are presented in Sect. 2. A greedy randomized adaptive search procedure (GRASP) is developed for its resolution in Sect. 3. A comparison between the design with rings (CmRSP) and the one with arbitrary 2-node-connected components (CmTNSSP) is presented in Sect. 4. Concluding remarks and trends for future work are discussed in Sect. 5.

2   Capacitated m Two-Node Survivable Star Problem

Inspired by fiber optics design, Martine Labbé et al. introduce the Ring Star Problem, or RSP for short [7]. The core is a ring, and the remaining terminals are pendant from the ring. The goal is to find the minimum-cost topology meeting the previous constraints, given different costs in the ring-connections and pendant-connections. A further generalization, the CmRSP, is introduced by Roberto Baldacci et al. [1]. The authors considered a depot and m rings, with the depot as the only common node. The main difference with the RSP is the presence of m rings instead of one. Both problems belong to the $\mathcal{NP}$-Hard class, since they represent a generalization of the Hamiltonian Tour [6]. Therefore, the CmRSP has been heuristically addressed in several opportunities [5,13].

We are given a simple graph $G = (V, E)$, a positive integer m and a tripartition $V = \{s\} \cup V_S \cup V_T$, being s the depot, V_S the optional Steiner nodes and V_T the terminal nodes. The source s has a capacity q_s, and there are two classes of connections with different costs: ring-connections are given by a cost-matrix $R = (r_{i,j})$, $v_i, v_j \in V$ and pendant-connections are given by another cost-matrix $C = (c_{i,j})$, $v_i \in V - \{s\}$, $v_j \in V_T$. In the CmRSP, the goal is to choose a minimum cost spanning subgraph $H = \cup_{i=1}^{m} C_{l_i} \cup S_i$, wherein the C_{l_i}s are cycles that only meet on the depot $s \in C_{l_i}$ and have a length l_i, and the S_is are pendant links from nodes belonging to C_{l_i}. The capacity constraint implies that $|S_i| + l_i \leq q_s$ for all $i \in \{1, \ldots, m\}$.

If we consider arbitrary 2-node-connected components instead of the rings C_{l_i}, we obtain the Capacitated m Two-Node Survivable Star Problem (CmTNSSP). The CmTNSSP also belongs to the $\mathcal{NP}$-Hard class of problems, since the design of one component ($m = 1$, $q_s = +\infty$, $V_S = \emptyset$) is the minimum-cost 2-node-connected spanning network problem (MW2NCSN), which is $\mathcal{NP}$-Hard. Monma et al. in their work [9] proved this for metric distances. They assigned a value 1 to the cost of the edges, then there exists a Hamiltonian cycle if and only if the minimum cost of MW2NCSN is equals to the number of nodes. Finally since "Hamiltonian Tour" belongs to Karp list [6] then MW2NCSN is $\mathcal{NP}$-Complete.

3   GRASP Resolution

Greedy Randomized Adaptive Search Procedure (GRASP) is a powerful multi-start or iterative process, with great success in telecommunications [12]. In GRASP, feasible solutions are produced in a first phase, while neighbor solutions are explored in a second phase. The best overall solution is returned as the result. There is a trade off between greediness (intensification) and randomization (diversification), by means of a restricted candidate list. For a comprehensive study of this metaheuristic see [10,11]. The main components of our particular GRASP design, namely Construction Phase and Local Search Phase, are depicted below.

3.1   Construction Phase

In this phase we build a feasible solution (see Algorithm 1). Each one of the m components are iteratively added to the solution, starting with one ring per component and then adding paths between two nodes of the same component until all terminal nodes are assigned. During the Construction Phase, no pendant links will be considered. The goal is to produce a feasible solution, despite the potential high cost of it (which will be reduced during Local Search Phase).

Let us consider an arbitrary instance for the CmTNSSP, a positive integer k and a maximum number of iterations $iter$. In order to define our construction phase, the following four functions will be used:

1 $Picking(m, G, R, iter)$: returns m terminal nodes $v_1, \ldots, v_m$, one for each component to build.
2 $Connecting(G, R, \hat{C}, s, node, k, non\_connected)$: connects each node v_i with the source-node s by k node-disjoint paths.
3 $ChooseTwo(\hat{C})$: randomly chooses 2 paths out of k using uniform distribution. At this stage one cycle per component is obtained.
4 $ConnectAllOthers(non\_connected, G, \hat{C})$: connects nodes that are not yet included in the construction with a component, adding a path between two nodes of such component.

Algorithm 1. Construction Phase

1: **input** G, R, V_T, s, k, m, $iter$
2: $G_{Sol} \leftarrow \emptyset$
3: $\hat{C} \leftarrow \emptyset$
4: $component\_nodes[m] \leftarrow \emptyset$ {Array with m empty positions}
5: $non\_connected \leftarrow V_T$
6: $\{v_1, \ldots, v_m\} \leftarrow Picking(m, G, R, iter)$
7: **for** i=1 **to** m **do**
8:     $node = Random(v_1, \ldots, v_m)$
9:     $\hat{C} \leftarrow Connecting(G, R, \hat{C}, s, node, k, non\_connected)$
10:     $C_i \leftarrow ChooseTwo(\hat{C})$
11:     $G_{Sol} \leftarrow G_{Sol} \cup C_i$
12:     $component\_nodes[i] \leftarrow component\_nodes[i] \cup C_i$
13:     $non\_connected \leftarrow non\_connected - C_i$
14: **end for**
15: $G_{Sol} \leftarrow G_{Sol} \cup ConnectAllOthers(non\_connected, G, \hat{C})$
16: **return** G_{Sol}

The previous functions will be run sequentially. *Picking* function returns a set of m terminal nodes by considering $iter$ sets of randomly chosen m nodes and returning the set with the greatest sum of costs of the edges determined by each pair of nodes (line 6).

Once the set $\{v_1, \ldots, v_m\}$ is obtained, *Connecting* function (line 9) connects node with the source-node s. Thus function is called for each *node* v_i which is selected randomly using the function *Random* (line 8). It applies Ramesh Bhandari's algorithm [3] in order to find the cheapest set of k node-disjoint paths between the depot and terminal node v_i.

Function *ChooseTwo* (line 10) just chooses uniformly at random two disjoint paths out of k from each component. Up to this point m rings that share the depot have been built.

Finally, in *ConnnectAllOthers* function (line 15), non-connected nodes are randomly chosen and iteratively added to the component with the lowest number of nodes. In this way, the capacity constraint is met during the construction phase, even though the cost could be high. Consider a non-connected node v and the (2-node-connected) component $\hat{C}$ (Fig. 2). All links that belong to other components will be deleted (i.e. only one component is treated at a time), and the costs of all links from $\hat{C}$ (grey edges) will temporarily be zero. We add an artificial node v' connected with all nodes from $\hat{C}$ using edges at zero cost (dotted edges). Bhandari's algorithm is applied in order to find the better k (or possibly less) node-disjoint paths between v and v' in the resulting network. Only two disjoint paths between v and v' will be uniformly chosen. Finally, we delete node v' and the resulting two paths that connect v with C are added to the solution.

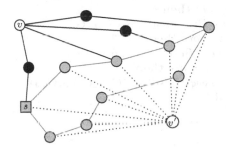

Fig. 2. Including node v into component $\hat{C}$.

3.2   Local Search Phase

The following operations fully determine neighborhood structures. A Variable Neighborhood Descent (VND [8]) scheme will be use to combine them.

- *Swapping*(G, R, C, V_T, G_{sol}): picks a random terminal node in G_{sol} and swaps it with its closest possible terminal node (the possibility means that the cost is decreased and the solution remains feasible),
- *ExtractInsert*(G, R, C, V_T, G_{sol}): extracts the links of a node, reconnects its neighbors and greedily inserts the node in G_{sol} (i.e., minimum cost insertion),
- *Crossing*(G, R, C, V_T, G_{sol}): picks two close terminal nodes from different components in G_{sol}, deletes one incident link from each node and reconnects components in the best manner,
- *BestPath*(G, R, C, V_T, G_{sol}) replaces any simple path with pendant nodes l in G_{sol} by the best of them (with the same endpoints), using an exact algorithm based on an Integer Linear Programming (ILP) model.

– $Best2NC(G, R, C, V_T, G_{sol})$: each cycle in G_{sol} is replaced by its best 2-node-connected component, using an exact algorithm based on ILP.

The full algorithm of GRASP-VND used in this paper, with the Construction phase and the sequence of local searches, is depicted in Algorithm 2.

Algorithm 2. Model of GRASP-VND used.

1: input G, R, C, V_T, s, k, m, $iter$, $grasp\_iter$, shk_iter,
2: **repeat**
3:    $G_{sol} \leftarrow Construction\_phase(G, R, V_T, s, k, m, iter)$
4:    $G_{iter} \leftarrow G_{sol}$
5:    **repeat**
6:       $improve=true$
7:       **while** $improve$ **do**
8:          $improve=Swapping(G, R, C, V_T, G_{sol})$
9:          **if** not $improve$ **then**
10:             $improve=ExtractInsert(G, R, C, V_T, G_{sol})$
11:             **if** not $improve$ **then**
12:                $improve=Crossing(G, R, C, V_T, G_{sol})$
13:                **if** not $improve$ **then**
14:                   $improve=BestPath(G, R, C, V_T, G_{sol})$
15:                   **if** not $improve$ **then**
16:                      $improve=Best2NC(G, R, C, V_T, G_{sol})$
17:                   end if
18:                end if
19:             end if
20:          end if
21:       **end while**
22:       **if** $\text{cost}(G_{sol}) < \text{cost}(G_{iter})$ **then** $G_{iter} \leftarrow G_{sol}$ **end if**
23:       $G_{sol} \leftarrow Shaking(C, R, V_T, G_{sol})$
24:    **until** shk_iter are reached
25:    **if** $\text{cost}(G_{iter}) < \text{cost}(G_{best})$ **then** $G_{best} \leftarrow G_{iter}$ **end if**
26: **until** $grasp\_iter$ are reached
27: **return** G_{best}

The first three local searches involve moves that have been usually applied to several network-based combinatorial optimization problems and they are explained in more detail in the thesis of Gabriel Bayá [2]. The remaining two local searches are detailed below.

Best Path with Pendants. This local search named *BestPath*, is based on an integer linear programming model. A preliminary concept is first introduced.

Definition 1 *Path with pendant nodes. Given an undirected graph $G = (V, E)$ we say that G is a path with pendant nodes which has endpoints a and $z \in V$ when there exists a path $l(a, z) \subseteq G$ that connects nodes a and z (which we call main path), and the following conditions are met:*

– G is a tree.
– All nodes that do not belong to l are directly connected to some node of l.

Given a feasible solution to the CmTNSSP we should identify all simple cycles that exist in each component and we should divide them in paths, adding their pendants nodes. Each path with pendants which has endpoints a and z is replaced by the best path with pendants with the same endpoints. This algorithm is based on an integer linear programming model.

We consider the following definitions:

Let $G = (V, E)$ be an undirected graph.

Let $\hat{T}$ be the set of terminal nodes of G.

Let $Adj(i)$ be the set of adjacent nodes to node $i \in V$ such:

$$Adj(i) = \{j \in V : (i, j) \in E\}$$

Let a and z be two distinguished terminal nodes such that $a \in \hat{T}$ and $z \in \hat{T}$.

Let $T = \hat{T} \setminus (\{a\} \cup \{z\})$ be the set of terminal nodes without a and z.

We define $R = \{r_{ij}\}_{i,j \in V}$ as the routing cost matrix of the graph, for each edge (i, j) which belongs to the main path $l(a, z)$.

Let us now define $C = \{c_{ij}\}_{i,j \in V}$ as the connection cost matrix of the graph, that is the cost of the edge (i, j) when one endpoint belongs to the main path and the other one does not belong to such path.

Let $W = V \setminus \hat{T}$ be the set of Steiner nodes. Let us now define the decision variables.

$$X_i = \begin{cases} 1 \text{ if node } i \in \hat{T} \text{ belongs to the main path} \\ 0 \text{ otherwise} \end{cases}$$

$$Y_i = \begin{cases} 1 \text{ if node } i \in T \text{ is a pendant node} \\ 0 \text{ otherwise} \end{cases}$$

$$z_{i,j} = \begin{cases} 1 \text{ if } i \in \hat{T} \text{ and } j \in V \text{ are connected, being } i \text{ a pendant node and} \\ \quad j \text{ j a node that belongs to the main path} \\ 0 \text{ otherwise} \end{cases}$$

$$x_{i,j} = \begin{cases} 1 \text{ if edge } (i, j) \text{ is used in the solution} \\ 0 \text{ otherwise} \end{cases}$$

$$w_{i,j} = \begin{cases} 1 \text{ if edge } (i, j) \text{ is a pendant edge and is used in the solution} \\ 0 \text{ otherwise} \end{cases}$$

$$y_{i,j}^{u,v} = \begin{cases} 1 \text{ if edge } (i, j) \text{ is used in the path that goes from node } u \text{ to node } v \\ 0 \text{ otherwise} \end{cases}$$

The integer linear programming model is defined as follows:

$$min(\sum_{i,j \in V} r_{ij}(x_{ij} - w_{ij}) + \sum_{i,j \in V} c_{ij}w_{ij}) \tag{1}$$

subject to:

$$X_i + Y_i = 1 \ \forall i \in T \tag{2}$$

$$X_i = 1 \ \forall i \in (\{a\} \cup \{z\}) \tag{3}$$

Equation 2 guarantees thay any terminal node which is not and endpoint either belongs to the main path or is pendant from the main path by a pendant edge, whereas constraint 3 ensures that the endpoints a and z belong exclusively to the main path.

$$z_{ij} \leq X_j \ \forall i \in T \ \forall j \in Adj(i) \tag{4}$$

$$Y_i = \sum_{j \in Adj(i)} z_{ij} \quad \forall i \in T \tag{5}$$

$$\sum_{j \in V} w_{i,j} \leq Y_i \ \forall i \in T \tag{6}$$

Constraint 4 implies that if i and j are connected and node i is a pendant node then node j belongs to the main path. Constraint 5 implies that if node i is pendant from the main path then it does so only by one edge. Constraint 6 ensures there is only one edge incident to a pendant node.

$$z_{i,j} = w_{i,j} \ \forall i \in T \quad j \in Adj(i) \tag{7}$$

$$\sum_{j \in Adj(i)} x_{i,j} \leq M(1 - Y_i) + 1 \ \forall i \in T, \ M \in \mathbb{Z}^+,$$
$$M \geq max(|Adj(i)|) \ \ i = 1 \cdots |V| \tag{8}$$

$$w_{i,j} \leq x_{i,j} \ \forall i \in T \quad j \in Adj(i) \tag{9}$$

Constraint 7 implies that if node i is pendant from node j then the edge (i, j) belongs to the solution. Inequality 8 constraints the degree of pendant nodes to 1 and it allows any other node of the main path to have any degree. Constraint 9 implies that if an edge is pendant then it belongs to the solution.

$$\sum_{j \in Adj[u]} y_{u,j}^{u,v} = 1 \ \forall u, v \in \hat{T}, u \neq v, \tag{10}$$

$$\sum_{i \in Adj[v]} y_{i,v}^{u,v} = 1 \ \forall u, v \in \hat{T}, v \neq u, \tag{11}$$

$$\sum_{i \in Adj[p]} y_{(i,p)}^{u,v} - \sum_{i \in Adj[p]} y_{p,i}^{u,v} \geq 0 \ \forall u, v \in \hat{T}, \ \forall p \in V \setminus u, v \tag{12}$$

$$y_{i,j}^{u,v} + y_{j,i}^{u,v} \leq x_{i,j} \ \forall u, v \in \hat{T}, u \neq v, \ \forall (i,j) \in E \tag{13}$$

Constraints 10 and 11 are simple connectivity constraints between nodes of any path (u, v). Constraint 12 is the balance equation of the internal nodes of the path. Constraint 13 guarantees that the path is edge-disjoint (i.e. a path which does not repeat any edge).

$$Y_i = 0 \ \forall i \in W \tag{14}$$

$$\left(\sum_{i \in Adj[j]} z_{i,j} + 2X_j - \sum_{i \in Adj[j]} x_{j,i} = 0 \right) \ \forall j \in W \tag{15}$$

$$\sum_{i \in Adj[j]} (z_{i,j} + z_{j,i}) + 2X_j - \sum_{i \in Adj[j]} x_{j,i} = 0 \ \forall j \in T \tag{16}$$

$$\sum_{i \in Adj[j]} (z_{i,j}) + X_j - \sum_{i \in Adj[j]} x_{j,i} = 0 \ \forall j \in (\{a\} \cup \{z\}) \tag{17}$$

In Eq. 14 it is ensured that Steiner nodes exclusively belong to the main path and constraints 15 to 17 are adjustment equations for Steiner, terminal and endpoint nodes. Algorithm 3 describes a local search which involves the replacement of a path with pendants by the best path with pendants. It begins by taking as input the graph G_{Sol}, which is a feasible solution of CmTNSSP. For each m components of G_{Sol} all of its cycles are counted, which are then identified and stored in the indexed list *all_cycles* (lines 3 and 4). Next, each of the cycles identified in the previous steps is treated, running the operations during **for** loop (lines 5 to 13) until all cycles are considered. Each cycle is divided into a certain number of paths of variable length. Next, we entered into a repetitive loop during the second **for** loop (lines 7 to 12), wherein each path obtained in the previous step is added with pendant nodes present in G_{Sol}, using the function **add_pendants** (line 8) obtaining a path with pendants P. In the next step, we generated the graph H induced by nodes of the path with pendants P with respect to the original graph G (line 9). Graph H thus generated is input of stage **best_pwp** which returns the best path with pendants (line 10). To accomplished this goal, **best_pwp** resolves the integer linear programming model depicted in (1)–(17). In line 11 P is replaced by P_{best} obtaining a better solution cost G_{best}. After processing all paths within each cycle, the best solution G_{best} is returned (line 15).

Algorithm 3. Best path with pendant nodes.

1: input G, R, C, V_T, G_{sol}
2: $G_{best} \leftarrow G_{sol}$
3: $q\_cycles = cycles\_count(G_{sol})$ {Numbers of cycles of G_{sol}}
4: $all\_cycles \leftarrow cycles(G_{sol})$ {Array with all cycles of G_{sol}}
5: **for** $(i = 1$ **to** $q\_cycles)$ **do**
6:    $paths=$**divide_into_paths**$(all\_cycles[i], q\_paths)$
7:    **for** $(j= 1$ **to** $q\_paths)$ **do**
8:       $P \leftarrow$ **add_pendants**$(G_{sol}, paths[j])$
9:       $H \leftarrow$ **induced_graph_path**(P, G)
10:      $P_{best} \leftarrow$ **best_pwp**(G_{sol}, P, R, C, H)
11:      $G_{best} \leftarrow G_{best}$ - P + P_{best}
12:   **end for**
13: **end for**
14: $improve=(\text{Cost}(G_{best}) < \text{Cost}(G_{sol}))$
15: **return** $improve$, G_{best}

Best 2-Connected Component. This local search named *Best2NC* is also based on integer linear programming. As in the previous local search, given a feasible solution to the problem, Algorithm 4 identifies all cycles that exist in each component. For each cycle we applied an exact algorithm getting the best replacement solution that changes a cycle by 2-node-connected topology.

As stated in Sect. 1, the best 2-node-connected solution covering a certain set of nodes is not necessarily a cycle, so this local search may include such topologies in our solution (see Fig. 1). This algorithm takes as input the induced sub-graph of the original graph with nodes of the cycle and some Steiner nodes, and returns the best 2-node-connected sub-graph, i.e. it can potentially change a cycle by a 2-node-connected topology if such change improves solution costs. In order to model this local search we used a particular case of **GSP** (General Steiner Problem) wherein connectivity of all its terminal nodes is two. We considered the following definitions:

Let $G = (V, E)$ be an undirected graph where V is the set of vertices and E is the set of edges of graph G.

Let $\hat{T}$ be the set of terminal nodes of graph G.

Define $R = \{r_{ij}\}_{i,j \in V}$ as the routing cost matrix, i.e. the costs when edge (i, j) belongs to the 2-node-connected structure of the component. In this local search, we only used such routing cost matrix since pendant nodes hitherto generated were not considered.

Model variables are defined below.

$$x_{i,j} = \begin{cases} 1 \text{ if edge } (i,j) \text{ is used in the solution} \\ 0 \text{ otherwise} \end{cases}$$

$$y_{i,j}^{u,v} = \begin{cases} 1 \text{ if edge } (i,j) \text{ is used in a path from node } u \text{ to } v \\ 0 \text{ otherwise} \end{cases}$$

Once the variables were specified, the integer linear programming model was defined as follows:

$$min(\sum_{i,j \in V} r_{ij}x_{ij}) \tag{18}$$

subject to:

$$\sum_{j \in Adj[u]} y_{u,j}^{u,v} = 2 \ \forall u,v \in \hat{T}, u \neq v, \tag{19}$$

$$\sum_{i \in Adj[v]} y_{i,v}^{u,v} = 2 \ \forall u,v \in \hat{T}, v \neq u, \tag{20}$$

$$\sum_{i \in Adj[p]} y_{i,p}^{u,v} - \sum_{i \in Adj[p]} y_{p,i}^{u,v} \geq 0 \ \forall u,v \in \hat{T}, \forall p \in V \setminus u,v \tag{21}$$

$$y_{i,j}^{u,v} + y_{j,i}^{u,v} \leq x_{i,j} \ \forall u,v \in \hat{T}, u \neq v, \forall (i,j) \in E \tag{22}$$

Algorithm 4. Best 2-node-connected component.

1: input G, G_{sol}
2: $G_{best} \leftarrow G_{sol}$
3: $q\_cycles = cycles\_count(G_{sol})$ {Number of cycles of G_{sol}}
4: $all\_cycles \leftarrow cycles(G_{sol})$ {Array with cycles of G_{sol}}
5: **for** $(i = 1 \text{ to } q\_cycles)$ **do**
6:    $best = \textbf{best\_component}(G_{best}, G, R, all\_cycles(i))$
7:    $G_{best} \leftarrow G_{best} - all\_cycles(i) + best$
8: **end for**
9: $improve = (\text{Cost}(G_{best}) < \text{Cost}(G_{sol}))$
10: **return** $improve$, G_{best}

Analogously to Algorithm 3, Algorithm 4 counts and identifies the cycles present in G_{sol} (lines 3 and 4). For each of these cycles the stage **best-component** (line 6) returns the best 2-node-connected structure and the cycle is replaced by the latter (performed in line 7). The function **best-component** resolves the integer linear programming model depicted in (18)–(22). It should be noted that neighbor solutions are feasible, so feasibility is preserved during the local search phase.

In order not to get stuck in a local optimum, a perturbation process takes place. Function *Shaking* randomly disconnects a proportion h of terminal nodes in the local optimal solution and reconnects them otherwise. *Shaking* is called whenever the previous five functions are stuck in a solution and do not have activity (i.e., they do not produce better solutions) as it can be seen in Algorithm 2.

4   Empirical Results

It must be observed that CmTNSSP is a relaxation of CmRSP. Therefore, the cost of feasible solutions for the CmTNSSP could be better than optimal values for the CmRSP. In order to highlight the main challenges of the new problem and the improvement offered by our GRASP methodology, we made a comparison with optimal solutions for the CmRSP, choosing instances developed by Roberto Baldacci, Mauro Dell' Amico and José Luis Salazar González in [1]. The authors considered instances from TSPLIB. Such instances are divided into two classes (A and B) using graphs with 26, 51, 76 and 101 nodes. Both classes have the same topology, but edge costs are different. In class A, the cost of each link equals the Euclidean distance $d_{i,j} = r_{i,j} = c_{i,j}$, while in class B, $r_{i,j} = \lceil 7d_{i,j} \rceil$ and $c_{i,j} = \lceil 3d_{i,j} \rceil$. We used $m \in \{3, 4, 5\}$. The GRASP algorithm has been executed using $k = 4$ for the restricted candidate list and $h = \lfloor 0, 3 \times |V_T| \rfloor$ for shaking, which were tuned with other smaller TSPLIB instances from Classes A and B. The heuristic was fully coded in C language using the CPLEX Callable Library to resolve integer linear programming models. Hardware where algorithms were run, consists of a computer with Intel I7 processor with 8 Gb. RAM and OS Fedora Core 20.

Tables 1 and 2 present a comparison between the optimal solution for the CmRSP (Z_1) found by Baldacci et al. [1] and the cost in CmTNSSP (Z_{best}) found by our proposed algorithm, for instances of Classes A and B, from a total of 90 instances tested; $\hat{Z}$ is the mean of 20 independent experiments for each instance and Z_2 is the best known value for CmRSP, recently published in [13]. The acronyms PN, CN, SN stand for the number of Pending Nodes, Connected Nodes and Steiner Nodes in the solutions, respectively.

The parameter *gap* is a measurement of our GRASP-VND effectiveness, and it is defined as follows:

$$gap = \frac{Z_{best} - Z_1}{Z_1}. \tag{23}$$

In particular, Tables 1 and 2 show the gaps with respect to Z_1, where negative values are highlighted in boldface. We can observe that for Class A, the objective

Table 1. Values found for Class A instances.

INSTANCE	$|T|$	Q	CN	PN	SN	$\hat{Z}$	Z_{best}	Z_1	Z_2	gap %	t(s)
A01-n026-m03	12	5	12	0	1	242	242	242	242	0,000	1.61
A02-n026-m04	12	4	12	0	1	261	261	261	261	0,000	0.97
A03-n026-m05	12	3	12	0	1	292	292	292	292	0,000	13.77
A04-n026-m03	18	7	18	0	0	301	301	301	301	0,000	34.29
A05-n026-m04	18	5	18	0	0	339	339	339	339	0,000	62.58
A06-n026-m05	18	4	18	0	0	375	375	375	375	0,000	2.67
A07-n026-m03	25	10	24	1	0	325	325	325	325	0,000	14.06
A08-n026-m04	25	7	25	0	0	362	362	362	362	0,000	3.99
A09-n026-m05	25	6	25	0	0	383	382	382	382	0,000	3.99
A10-n051-m03	12	5	12	0	0	242	242	242	242	0,000	20.09
A11-n051-m04	12	4	12	0	3	261	261	261	261	0,000	6.42
A12-n051-m05	12	3	11	1	2	286	286	286	286	0,000	37.69
A13-n051-m03	25	10	22	3	3	322	322	322	322	0,000	130.85
A14-n051-m04	25	7	24	1	1	360	360	360	360	0,000	49.75
A15-n051-m05	25	6	23	2	2	379	379	379	379	0,000	117.67
A16-n051-m03	37	14	33	4	1	373	373	373	373	0,000	296.60
A17-n051-m04	37	11	33	4	1	405	405	405	405	0,000	80.49
A18-n051-m05	37	9	33	4	1	434	432	432	432	0,000	2720.60
A19-n051-m03	50	19	45	5	0	461	458	458	458	0,000	1674.86
A20-n051-m04	50	14	48	2	0	492	490	490	490	0,000	3429.11
A21-n051-m05	50	12	43	7	0	521	520	520	520	0,000	6338.64
A22-n070-m03	18	7	17	1	5	332	330	330	330	0,000	36.13
A23-n076-m04	18	5	15	3	7	385	385	385	385	0,000	112.97
A24-n076-m05	18	4	17	1	4	448	448	448	448	0,000	109.91
A25-n076-m03	37	14	35	2	2	403	403	402	402	0,249	3624.35
A26-n076-m04	**37**	**11**	**40336**	**1**	**3**	**458**	**456**	**460**	**457**	**-0,870**	**7200.00**
A27-n076-m05	37	9	36	1	4	483	483	479	479	0,835	7200.00
A28-n076-m03	56	21	48	8	1	474	474	471	471	0,637	7200.00
A29-n076-m04	**56**	**16**	**49**	**7**	**1**	**522**	**519**	**523**	**519**	**-0,765**	**7200,00**
A30-n076-m05	56	13	50	6	2	555	547	545	545	0,367	7200.00
A31-n076-m03	75	28	71	4	0	572	571	564	564	1,241	7200.00
A32-n076-m04	75	21	73	2	0	614	611	606	602	1,808	7200.00
A33-n076-m05	**75**	**17**	**68**	**7**	**0**	**657**	**651**	**654**	**640**	**-0,459**	**7200.00**
A34-n101-m03	25	10	21	4	7	370	363	363	363	0,000	199.27
A35-n101-m04	25	7	21	4	9	417	415	415	415	0,000	1023.84
A36-n101-m05	25	6	22	3	9	453	448	448	448	0,000	1264.62
A37-n101-m03	50	19	46	4	8	503	500	500	500	0,000	4020.65
A38-n101-m04	50	14	47	3	6	545	538	532	528	1,128	7200.00
A39-n101-m05	50	12	46	4	5	578	573	568	567	0,880	7200.00
A40-n101-m03	75	28	69	6	5	616	613	595	595	3,025	7200.00
A41-n101-m04	75	21	73	2	1	656	651	625	623	4,160	7200.00
A42-n101-m04	75	17	70	5	2	680	677	662	657	2,266	7200.00
A43-n101-m03	100	38	84	16	0	665	662	646	646	2,477	7200.00
A44-n101-m04	100	28	87	13	0	684	680	680	679	0,000	7200.00
A45-n101-m05	100	23	84	16	0	722	713	700	700	1,857	7200.00

Table 2. Values found for Class B instances.

INSTANCE	$|T|$	Q	CN	PN	SN	$\hat{Z}$	Z_{best}	Z_1	Z_2	gap %	t(s)
B01-n026-m03	12	5	11	1	1	1684	1684	1684	1684	0,000	3.09
B02-n026-m04	12	4	12	0	1	1827	1827	1827	1827	0,000	1.09
B03-n026-m05	12	3	11	1	2	2041	2041	2041	2041	0,000	10.68
B04-n026-m03	18	7	17	1	1	2104	2104	2104	2104	0,000	24.90
B05-n026-m04	18	5	17	1	1	2370	2370	2370	2370	0,000	78.21
B06-n026-m05	18	4	17	1	2	2615	2615	2615	2615	0,000	47.01
B07-n026-m03	25	10	24	1	0	2251	2251	2251	2251	0,000	35.13
B08-n026-m04	25	7	24	1	0	2512	2510	2510	2510	0,000	51.65
B09-n026-m05	25	6	25	0	0	2677	2674	2674	2674	0,000	150.31
B10-n051-m03	12	5	10	2	2	1681	1681	1681	1681	0,000	2035.19
B11-n051-m04	12	4	10	2	3	1821	1821	1821	1821	0,000	49.26
B12-n051-m05	12	3	10	2	2	1976	1975	1972	1972	0,152	930.42
B13-n051-m03	25	10	21	4	3	2176	2176	2176	2176	0,000	1724.28
B14-n051-m04	25	7	22	3	3	2471	2470	2470	2470	0,000	626.97
B15-n051-m05	25	6	21	4	4	2596	2579	2579	2579	0,000	92.66
B16-n051-m03	37	14	29	8	2	2498	2490	2490	2490	0,000	3699.45
B17-n051-m04	37	11	29	8	2	2747	2735	2721	2721	0,515	3605.47
B18-n051-m05	37	9	32	5	2	2931	2908	2908	2908	0,000	197.51
B19-n051-m03	50	19	39	11	0	3028	3015	3015	3015	0,000	871.33
B20-n051-m04	50	14	39	11	0	3284	3267	3260	3260	0,215	7200,00
B21-n051-m05	50	12	38	12	0	3426	3404	3404	3404	0,000	3773.22
B22-n076-m03	18	7	15	3	4	2258	2253	2253	2253	0,000	186.10
B23-n076-m04	18	5	13	5	8	2661	2620	2620	2620	0,000	90.78
B24-n076-m05	18	4	15	3	9	3142	3155	3059	3059	3,138	7200,00
B25-n076-m03	37	14	32	5	6	2747	2731	2720	2720	0,404	7200,00
B26-n076-m04	**37**	**11**	**34**	**3**	**4**	**3142**	**3134**	**3138**	**3100**	**-0,127**	**7200.00**
B27-n076-m05	37	9	36	1	3	3327	3329	3311	3284	0,544	7217.19
B28-n076-m03	**56**	**21**	**40**	**16**	**4**	**3060**	**3044**	**3088**	**3044**	**-1,425**	**7200.00**
B29-n076-m04	**56**	**16**	**44**	**12**	**2**	**3448**	**3439**	**3447**	**3415**	**-0,232**	**7200.00**
B30-n076-m05	**56**	**13**	**44**	**12**	**2**	**3676**	**3635**	**3648**	**3632**	**-0,356**	**3797.03**
B31-n076-m03	**75**	**28**	**55**	**20**	**0**	**3742**	**3724**	**3740**	**3652**	**-0,428**	**2112.23**
B32-n076-m04	75	21	57	18	0	4102	4096	4026	3964	1,739	7200,00
B33-n076-m05	75	17	58	17	0	4512	4489	4288	4217	4,688	7200,00
B34-n101-m04	25	7	19	6	9	2452	2445	2434	2434	0,369	7200,00
B35-n101-m04	25	7	19	6	6	2804	2795	2782	2782	0,467	7200,00
B36-n101-m05	25	6	18	7	4	3015	3009	3009	3009	0,000	597.71
B37-n101-m03	**50**	**19**	**40**	**10**	**8**	**3338**	**3331**	**3332**	**3322**	**-0,030**	**7200,00**
B38-n101-m04	50	14	38	12	8	3616	3560	3533	3533	0,764	7200,00
B39-n101-m05	50	12	41	9	8	3895	3873	3872	3834	0,026	7200,00
B40-n101-m03	75	28	68	7	5	3958	3931	3923	3887	0,204	7200,00
B41-n101-m04	75	21	68	7	6	4345	4332	4125	4082	5,018	7200,00
B42-n101-m05	75	17	69	6	6	4556	4494	4458	4358	0,808	7200,00
B43-n101-m03	100	38	96	4	0	4413	4403	4110	4110	7,129	7200,00
B44-n101-m04	100	28	95	5	0	4560	4526	4506	4355	0,444	7200,00
B45-n101-m05	100	23	96	4	0	4645	4639	4632	4565	0,151	7200,00

value obtained for the CmTNSSP is lower than its counterpart for CmRSP in 3 instances and equal in 29 instances out of 45 with an average gap of 0.741 %. For Class B, the same fact can be observed in 6 and 21 respectively out of 45 instances with an average gap of 0.890 %, suggesting that the cost structure of this class promotes the application of CmTNSSP solutions.

5   Concluding Remarks and Future Work

The Capacitated m Two-Node Survivable Star Problem (CmTNSSP) has been introduced. As far as we are know, it has not been studied in prior literature. The need for redundancy and cheaper costs in network deployment is remarkable. Inspired by predictions from Clyde Monma and the previous CmRSP, we proposed an alternative problem, where rings are replaced by arbitrary 2-node-connected components. Both problems are computationally intractable. Therefore, heuristics are suitable for large case scenarios. As a corollary, the CmTNSSP has been heuristically addressed, following a hybrid GRASP metaheuristic that combines the resolutions of ILP models. The resulting topology could be cheaper than the one offered by the CmRSP but 2-node-connected as well. As a future work, we wish to apply these techniques to the design of real life networks. Indeed, optimal solutions for the CmTNSSP could by equally robust and more cost-effective than that of CmRSP.

References

1. Baldacci, R., Dell'Amico, M., González, J.J.S.: The capacitated m-ring-star problem. Oper. Res. **55**(6), 1147–1162 (2007)
2. Mantani, G.B.: Diseño Topológico de Redes. Caso de Estudio: Capacitated m Two-Node Survivable Star Problem. Master's thesis, Universidad de la República. Pedeciba Informática, Montevideo, Uruguay (2014)
3. Bhandari, R.: Optimal physical diversity algorithms and survivable networks. In: 1997 Second IEEE Symposium on Computers and Communications, pp. 433–441 (1997)
4. Canale, E., Monzón, P., Robledo, F.: Global synchronization properties for different classes of underlying interconnection graphs for Kuramoto coupled oscillators. In: Lee, Y., Kim, T., Fang, W., Ślęzak, D. (eds.) FGIT 2009. LNCS, vol. 5899, pp. 104–111. Springer, Heidelberg (2009)
5. Hoshino, E.A., de Souza, C.C.: A branch-and-cut-and-price approach for the capacitated m-ring-star problem. Discrete Appl. Math. **160**(18), 2728–2741 (2012)
6. Karp, R.M.: Reducibility among combinatorial problems. In: Miller, R.E., Thatcher, J.W. (eds.) Complexity of Computer Computations, pp. 85–103. Plenum Press, New York (1972)
7. Labbé, M., Laporte, G., Martín, I.R., González, J.J.S.: The ring star problem: polyhedral analysis and exact algorithm. Networks **43**(3), 177–189 (2004)
8. Mladenovic, N., Hansen, P.: Variable neighborhood search. Comput. Oper. Res. **24**(11), 1097–1100 (1997)
9. Monma, C., Munson, B.S., Pulleyblank, W.R.: Minimum-weight two-connected spanning networks. Math. Program. **46**(1–3), 153–171 (1990)

10. Resende, M., Ribeiro, C.: Greedy randomized adaptive search procedures. In: Glover, F., Kochenberger, G. (eds.) Handbook of Methaheuristics. Kluwer Academic Publishers, Norwell (2003)
11. Resende, M., Ribeiro, C.: GRASP: Greedy randomized adaptive search procedures. In: Burke, E.K., Kendall, G. (eds.) Search Methodologies, pp. 287–312. Springer, US (2014)
12. Robledo, F.: GRASP heuristics for Wide Area Network design. Ph.D. thesis, INRIA/IRISA, Université de Rennes I, Rennes, France (2005)
13. Zizhen Zhang, H., Qin, A.L.: A memetic algorithm for the capacitated m-ring-star problem. Appl. Intell. **40**(2), 305–321 (2014)

Robust Berth Allocation Using a Hybrid Approach Combining Branch-and-Cut and the Genetic Algorithm

Ghazwan Alsoufi[(✉)], Xinan Yang, and Abdellah Salhi

Department of Mathematical Sciences, University of Essex, Colchester, UK
{ghmals,xyangk,as}@essex.ac.uk

Abstract. Seaside operations at container ports often suffer from uncertainty due to events such as the variation in arrival and/or processing time of vessels, weather conditions and others. Finding a robust plan which can accommodate this uncertainty is therefore desirable to port operators. This paper suggests ways to generate robust berth allocation plans in container terminals. The problem is first formulated as a mixed-integer programming model whose main objective is to minimize the total tardiness of vessel departure time. It is then solved exactly and approximately. Experimental results show that only small instances of the proposed model can be solved exactly. To handle large instances in reasonable times, the Genetic Algorithm (GA) is used. However, it does not guarantee optimality and often the approximate solutions returned are of low quality. A hybrid meta-heuristic which combines Branch-and-Cut (B&C) as implemented in CPLEX, with the GA as we implement it here, is therefore suggested. This hybrid method retains the accuracy of Branch-and-Cut and the efficiency of GA. Numerical results obtained with the three approaches on a representative set of instances of the problem are reported.

Keywords: Container terminals · Berth allocation problem · Robustness · Genetic algorithm · Hybrid metaheuristic

1 Introduction

Container transportation is at the heart of the import and export of goods. The efficiency with which containers are loaded/unloded from vessels and ships are handled/processed, is affected by the time and place at which a vessel is moored along a berth. Finding this time and place forms the well known Berth Allocation Problem or BAP. The deterministic form of BAP is the one often solved in practice. However, a number of uncertain factors and unexpected events such as the deviation in arrival time and in the operations time of the initial baseline schedule affect BAP. If these uncertainties are ignored when drawing berth allocation plans, last minute scrambling and changes of plans, may ensue. Port managers, therefore, try protect the initial baseline schedule from the adverse

© Springer International Publishing Switzerland 2016
M.J. Blesa et al. (Eds.): HM 2016, LNCS 9668, pp. 187–201, 2016.
DOI: 10.1007/978-3-319-39636-1_14

effects of possible disruptions. Hence, there is a need for robust schedules or schedules which are not perturbed by variance in key parameters. This is the aim of this study.

We consider BAP under continuous wharf set-up with the main aim being to find the optimal berthing time and berthing position for each vessel arriving at the container terminal. We develop a mathematical model of the optimisation type the objective function of which consists of the operations cost related to where the vessel is to be moored and a penalty cost to pay for the potential delay in the departure time. Unloaded containers are placed temporarily at a pre-allocated yard storage space. How far the berthing position is from the yard storage space impacts on the time required to move a container and therefore the efficiency of processing a vessel. As a result, a "best berthing position" is nominated for every single incoming vessel, which is the closest berth position to the allocated yard storage space. If the best berthing position cannot be guaranteed, a "movement cost" for the additional time and effort for moving the container will have to be paid.

In this paper we develop a robust model by inserting a time buffers between the exact processing time of vessels, to add more flexibility to the final berth allocation plan. Unlike in previous studies where people mainly make the time buffers depend on the shipping line's reputation of punctuality, here we consider the expected processing time of a vessel when inserting time buffers between vessels. For example, if there are two vessels that are owned by the same shipping line with one carrying 50 containers to unload and the other 500, then obviously the one with fewer container (smaller workload; shorter expected processing time) has much less opportunity to run late compared with the one with higher workload. For this reason, a new weight for each vessel is added which represents the proportion of the processing time of a vessel over the sum of the processing time of all vessels under consideration. In practice, this weight can be seen as a measure of the possibility of having adverse events (faulty machinery such as quay cranes, trucks, yard cranes etc.) occurring during the processing time of the vessel.

The model suggested here is suitable for robust berth allocation in realistic situations. The contribution of this paper is three fold:

1. formulation of a mathematical model with more advanced weight parameters for robustness;
2. solution of realistic instances of the problem using an adapted variant of the genetic algorithm;
3. design and application of a hybrid meta-heuristic to improve performance on realistic instances of the problem.

The rest of the paper is organised as follows. Section 2 is a literature review of BAP. Section 3 presents the proposed mathematical model. In Sect. 4, a variant of the genetic algorithm is applied to the Robust BAP or RBAP; this variant is then combined with Branch-and-Cut to provide more effective solution approach.

Section 5 records the comparison results between B&C as implemented in CPLEX, GA and the hybrid meta-heuristic. Finally, Sect. 6 gives the conclusion and further work.

2 Literature Review

BAP is the problem of allocating berthing time and berthing position to vessels arriving at the port. There are several versions of BAP which are different in terms of wharf type (discrete or continuous), main objective (minimizing delay or operations cost).

The BAP with continuous wharf is the normal setting in most modern container ports as it offers more flexibility. Our study assumes also a continuous wharf setting. Under similar problem settings, Li et al. [11] formulated the problem in a "multiple-job-on-one-processor" pattern, where multiple jobs refer to several vessels and one processor refers to a single berth. A small vessel moored at berth may share the berth with other vessels if the total length does not exceed the length of the wharf. Lim [12] formulated a BAP and considered that the berthing time is equal to arrival time for each vessel and by solving his model, a berthing position is found which minimizes the maximum quay length required to serve vessels in accordance with the schedule. Moon [13] proposed a mixed integer linear programming model, whose objective is to minimize the tardiness of vessels. In order to overcome the difficulty of computation of the mixed integer program, the author designed a Lagrangean relaxation model which is solved by a sub-gradient optimization algorithm. Goh et al. [5] discussed the methods of modelling BAP and proposed several approaches to solve it such as the Randomized Local Search (RLS), the Genetic Algorithm (GA) and Tabu Search (TS). Kim et al. [17] formulated BAP as a mixed integer program and applied a simulated annealing algorithm to find near optimal solutions. Guan et al. [6] developed two inter-related BAP mathematical models the objective functions of which are to minimize the total weighted finishing time of vessels. A tree search procedure was used to solve the first model. This provide a good lower bound which then sped up the tree search procedure in the second one.

Imai et al. [8] addressed BAP in a multi-user container terminal and introduced a nonlinear programming model to represent it. The authors presented a heuristic for BAP in continuous wharf. In this paper, they addressed the preference of the flexible berth layout which has become very important especially in busy hub ports where ships of various sizes dock. Wang et al. [22] transformed BAP into a multiple stage decision making problem and a new multiple stage search method, namely the stochastic beam search algorithm was used to solve it. Lee [10] proposed a neighborhood-search based heuristic to determine the berthing time and space for each incoming vessel to the continuous berth stretch. In their method, the First-Come-First-Served rule, clearance distance between vessels and the possibility of vessel shifting were considered. Lee et al. [9] studied the continuous and dynamic BAP in order to minimize the total weighted flow time. The authors follow the mathematical model of Guan et al. [6]. Two versions of the Greedy Randomized Adaptive Search Procedures (GRASP) heuristic were developed to find near optimal solutions. Ganji et al. [4] proposed a genetic algorithm for large scale BAPs which we adapted ourselves in this paper to find optimal or near optimal solutions. Cheong et al. [2] solved BAP by using multi-objective optimization in order to minimize concurrently the three objectives of makespan, waiting time, and degree of deviation from a predetermined priority schedule. These three objectives represent the interests of both port and ship operators. Unlike most existing approaches in single-objective optimization, in multi-objective evolutionary algorithm (MOEA) that incorporates the concept of Pareto optimality is proposed for solving the multi-objective BAP.

In addition to what has been mentioned above, other more recent studies dealt with BAP under uncertainty. Moorthy et al. [14] studied how to design a robust berth template for the special requirements of transshipment hubs. Zhen et al. [25] studied the berth allocation problem under uncertain arrival time or operation time of vessels and proposed a two-stage stochastic programming model. A meta-heuristic approach is proposed for solving the above problem in large-scale realistic environments. Xu et al. [23] studied a robust berth allocation problem, which explicitly considers the uncertainty of vessel arrival and handling time. Time buffers are inserted between the vessels occupying the same berthing location to give room for uncertain delays. Using total departure delay of vessels as the service measure and the length of time buffer as the robustness measure, the authors formulated RBAP to balance the service level and plan robustness. Based on the properties of the optimal solution, the researchers developed a robust berth scheduling algorithm that integrates simulated annealing and Branch-and-Bound algorithms. This work considers tardiness only in its objective rather than the preferred berthing location which weakens the connections with the yard management problem. It also uses a constant time buffer which is independent of the reputation of the shipping line and the expected processing time. Zhen et al. [24] proposed a proactive strategy for making robust baseline schedules of BAP. A bi-objective nonlinear optimization model for minimizing cost and maximizing robustness is introduced. However, the model is not complete as some constraints used in the model depended on its solution. As a result the author did not even attempt to solve the optimization model. To evaluate their solution for large scale instances, they used extensive simulations and justified their solution against some practical criteria.

3   Mathematical Model

In this section we describe a mixed integer programming model for berth allocation in container terminals with continuous berths. The novelty of this model is that, unlike previous berth allocation/planning models, as mentioned earlier, its solutions provide robust berth plans. By this, we mean that the solutions mitigate the uncertainty that is often experienced at the level of such maritime facilities. Note that the solution to the model is in terms of optimum berthing time, berthing position and, the optimum time buffer between the berthing times of vessels. In the following a number of aspects of this model are given.

3.1 Assumptions

A number of assumptions have to be satisfied in order to apply the model. However, they are realistic. Given a set V of vessels referred to as v, v_i or v_j for any vessel, the ith or the j_{th} vessels respectively, we assume that:

1. every segment of the continuous wharf can handle only one vessel at a time;
2. there is a safety distance between each pair of adjacent vessels;
3. once the processing of vessel starts the vessel will only leave after its processing has finished;
4. a vessel can be handled at any place of the wharf depending on its arrival time and the availability of wharf space.

3.2 Parameters

W	Length of the wharf.
$- A_v$	Estimated arrival time for vessel v.
$- d_v$	Requested departure time for vessel v.
$- h_v$	Estimated operation/processing time to handle vessel v.
$- L_v$	Length of vessel v.
$- b_v$	Desired berthing position of vessel v; it is determined by the position of yard storage areas allocated to vessel v.
$- C_{1v}$	Tardiness cost of vessel v.
C_{2v}	Distance cost of vessel v for mooring away from b_v.
$- IN_v$	Instability in arrival time of vessel v.
$- PP_v$	Proportion of the processing time of vessel v over the processing time of all vessels.
$- R_v$	$IN_v + PP_v$ of the vessel v.
$- \lambda_v$	Length of time buffer.
$- M$	Arbitrary large positive number.

3.3 Binary Decision Variables

$$\delta_{v_i v_j} = \begin{cases} 1 & \text{if the processing of vessel } v_j \text{ starts later} \\ & \text{than the finishing time of vessel } v_i. \\ 0 & \text{otherwise} \end{cases}$$

$$\sigma_{v_i v_j} = \begin{cases} 1 & \text{if the vessel } v_j \text{ is located below the vessel } v_i \text{ in the berth.} \\ 0 & \text{otherwise.} \end{cases}$$

$$\xi_{v_i v_j} = \begin{cases} 1 & \text{if the vessel } v_j \text{ occupies part of the berthing position of vessel } v_i. \\ 0 & \text{otherwise.} \end{cases}$$

$$\zeta_{v_i v_j} = \begin{cases} 1 & \text{if the vessel } v_j \text{ occupies part of the berthing position} \\ & \text{of vessel } v_i \text{ and starts later than } v_i. \\ 0 & \text{otherwise.} \end{cases}$$

3.4   Other Decision Variables

- T_v Berthing time of vessel v.
- P_v Berthing position of vessel v.
- $\tau_{v_i v_j}$ Time buffer between vessels v_i and v_j.
- θ_v Minimum time buffer between vessel v_i and the other vessels.

3.5   The Mathematical Model

The explicit model we built for berth allocation is as follows.

$$\min \sum_{v=1}^{V} C_{1v}(T_v + h_v - d_v)^+ + \sum_{v=1}^{V} C_{2v}|P_v - b_v| \tag{1}$$

s.t

$$T_{v_i} + h_{v_i} + R_{v_i}\lambda_{v_i} - \theta_{v_i} \leq T_{v_j} + M(1 - \delta_{v_i v_j}) \qquad \forall v_i, v_j; v_i \neq v_j \tag{2}$$

$$P_{v_i} + L_{v_i} \leq P_{v_j} + M(1 - \sigma_{v_i v_j}) \qquad \forall v_i, v_j; v_i \neq v_j \tag{3}$$

$$\sigma_{v_i v_j} + \sigma_{v_j v_i} + \delta_{v_i v_j} + \delta_{v_j v_i} \geq 1 \qquad \forall v_i, v_j; v_i \neq v_j \tag{4}$$

$$T_v \geq A_v \qquad \forall v \tag{5}$$

$$0 \leq P_v + L_v \leq W \qquad \forall v \tag{6}$$

$$\xi_{v_i v_j} = 1 - (\sigma_{v_i v_j} + \sigma_{v_j v_i}) \qquad \forall v_i, v_j; v_i \neq v_j \tag{7}$$

$$\zeta_{v_i v_j} \geq \delta_{v_i v_j} + \xi_{v_j v_i} - 1 \qquad \forall v_i, v_j; v_i \neq v_j \tag{8}$$

$$\tau_{v_i v_j} \leq M(1 - \zeta_{v_j v_i}) + T_{v_j} - T_{v_i} - h_{v_i} \qquad \forall v_i, v_j; v_i \neq v_j \tag{9}$$

$$\theta_{v_i} \leq \tau_{ij} + M(1 - \zeta_{v_j v_i}) \qquad \forall v_i, v_j; v_i \neq v_j \tag{10}$$

$$\delta_{v_i v_j}, \sigma_{v_i v_j}, \zeta_{v_i v_j}, \xi_{v_i v_j} \in \{0, 1\} \tag{11}$$

$$P_{v_i}, T_{v_i}, \tau_{v_i v_j}, \theta_{v_i v_j} \geq 0 \tag{12}$$

Note that the objective function to minimise is made up of the cost of tardiness which is represented by the term $\sum_{v=1}^{V} C_{1v}(T_v + h_v - d_v)$ and the cost of the vessel being moored at an undesired berthing position represented by the term $\sum_{v=1}^{V} C_{2v}|p_v - b_v|$.

Constraints (2) define $\delta_{v_i v_j}$ such that $\delta_{v_i v_j} = 0$ or 1 if the finishing time of vessel i is less than or equal to the berthing time of vessel j; 0 if the finishing time of vessel i is greater than the berthing time of vessel j. λ is the time buffer between the finishing time of vessel i and the berthing time of vessel j. The decision maker at the port will select the value of λ according to his preferred aspect of the final plan, cost-effective or robust. A small λ means a cost-effective berth plan, while a large λ means a robust plan. The length of the time buffer between the vessel and its immediately posterior vessels can be computed by multiplying the associated value of the robustness parameter, R_{v_i}, with λ. The constraints set (2) works in such a way that, if the value of the product $R_{v_i}\lambda_{v_i}$ is less than the existing gap between these two vessels, θ_{v_i}, then we do not need to add any time buffer. Otherwise we need to increase the gap between these two vessels; the value of time buffer is the difference between $R_{v_i}\lambda_{v_i}$ and θ_{v_i}.

Constraints (3) guarantee that $\sigma_{v_i v_j} = 0$ or 1 depending on whether the berthing position of vessel i plus the length of vessel i is less than or equal to the berthing position of vessel j or 0 otherwise. Constraints (4) ensure that overlaps amongst vessels do not occur in the 2-dimensional space (time and location) depending on the values of $\delta_{v_i v_j}$ and $\sigma_{v_i v_j}$. Constraints (5) guarantee that the vessels cannot moor before their arrivals. Constraints (6) imply that the berthing position plus the length of the vessel cannot exceed the range of the wharf. Constraints (7) make variable $\xi_{v_i v_j}$ take value 1 if the vessel v_j occupies part of the berthing position of vessel v_i. Constraints (8) make variable $\zeta_{v_i v_j}$ take value 1 if vessel v_j occupies part of the berthing position of vessel v_i and starts later than vessel v_i. Constraints (9 and 10) compute θ_{v_i} for vessel v, which is the gap between vessel v_i and its immediately preceding vessel.

3.6 Numerical Example

Here, we apply B&C in CPLEX to a small instance of the above model involving six vessels. The rest of the data is given in Table 1.

Table 1. Example input data to CPLEX

Arrival time	0	35	40	85	100	110
Estimated processing time	30	30	40	20	20	30
Departure time	20	55	70	95	110	130
Length of vessel	70	50	40	70	35	45
Preferred position	20	50	0	30	0	70
Tardiness cost	1	1	1	1	1	1
Distance cost	1	1	1	1	1	1
Instability in arrival	0.8	0.2	0.8	0.9	0.7	0.5
proportion of the processing time	0.17	0.17	0.23	0.11	0.11	0.17

The solution returned by CPLEX consists of the berthing time and position of each vessel with appropriate time buffer between them to give robustness to the plan represented in Figs. 1, 2, 3 and 4. Clearly, the decision maker can use different values of λ to affect the property of the final solution sought.

4   Solution by the Genetic Algorithm

In order to solve large scale problems in reasonable time and to overcome the difficulty of the Branch-and-Cut (B&C) method, a Genetic Algorithm (GA) is used to find optimal or near optimal solutions. The choice of GA here as the approximate solution approach is dictated by it being well established and reliable. Potential users may adopt it readily. It was proposed by Holland [7]. It

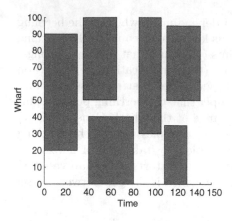

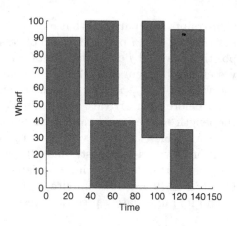

Fig. 1. Berthing plan when $\lambda = 5$; Obj.Fun = 83 (0 35 40 85 108 110 20 50 0 30 0 50)

Fig. 2. Berthing plan when $\lambda = 10$; Obj.Fun = 90 (0 35 40 86 112 112 20 50 0 30 0 50)

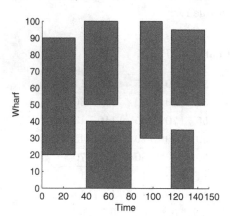

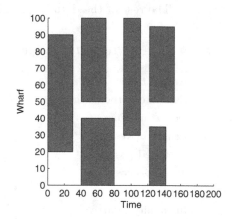

Fig. 3. Berthing plan when $\lambda = 15$; Obj.Fun = 103 (0 38 40 88 116 116 20 50 0 30 0 50)

Fig. 4. Berthing plan when $\lambda = 20$; Obj.Fun = 120 (0 40 40 91 122 122 20 50 0 30 0 50)

is an adaptive heuristic method based on natural evolution ideas. It repeatedly modifies a population of solutions, selecting individuals from the current population to be parents which then produce the children of the next generation. Over successive generations the population evolves towards an optimal solution. The processes involved in the GA are outlined below [15].

1. Generate an initial population: individual solutions (chromosomes) are created randomly to form an initial population.
2. Evaluate the fitness of each individual: choice of parents of new individuals for the new generations is biased toward individuals with good fitness values.

3. Create children: breed new individuals using genetic operators such as crossover, mutation and reproduction.
4. Generate new population: replace the worst individuals in the population with better new ones.
5. Stopping: The process is repeated until stopping criteria are met; these may include the specified maximum number of generations or time limit, high enough fitness etc.

The implementation of GA requires a good representation of individuals and a good fitness function. A good initial population helps the search. The GA which has been proposed by Ganji et al. [4], will be used with some tuning to find optimal or near optimal solutions.

4.1 Representation

A solution or chromosome is a strand of genes made of two parts of equal lengths. The first part represents berthing times (T_v) of vessels while the second represents berthing positions (P_v). The chromosomes are character rather than binary strings. See Fig. 5 for the chromosome representation of a 3-ship problem. The example shows a solution with ships 1, 2, and 3 being serviced in that order at times 42, 37, and 65 along the time axis, and at positions 213, 185, and 370 along the quay axis.

Chromosome	42	37	65	213	185	370
$T_i \& P_i$	T_1	T_2	T_3	P_1	P_2	P_3

Fig. 5. Representation

4.2 Initial Population

The population sizes used are a 100 or 500 individuals depending on the size of the problem instance. The populations are generated randomly but within the feasible solution set defined by the constraints (5), (6) and (12). For each solution $((T_v), (P_v))$, one can compute the handling times (C_v), and the completion times of handling (F_v) given the input parameters $h_v, R_{v_i}, \lambda_{v_i}$ and the decision variable θ_{v_i} as follows.

Handling or processing time: $C_v = h_v + R_{v_i}\lambda_{v_i} - \theta_{v_i}$.

Completion or finishing time: $F_v = T_v + C_v$.

4.3   Fitness Function with a Penalty Term

Each generated random solution is checked against constraints (2), (3) and (4) to see that there is no overlapping of ships in time and place dimensions. A solution that satisfies these constraints is accepted. Otherwise it is accepted after addition of a penalty term to its objective function value Z. The objective function value and the penalty term when used form the fitness of that solution. The penalty term in the fitness function gradually removes infeasible solutions from the next generations. This term is computed as $\gamma_{ij} = (A_{ij} \times B_{ij})^{0.5}$ based on the area of overlap in time and space between vessels i and j where

$$A_{ij} := Max(\frac{L_i + L_j}{2} - |\frac{P_i + P_i + Li}{2} - \frac{P_j + P_j + L_j}{2}|, 0), \text{and}$$

$$B_{ij} := Max(\frac{C_i + C_j}{2} - |\frac{t_i^B + t_i^F}{2} - \frac{t_j^B + t_j^F}{2}|, 0).$$

A_{ij} and B_{ij} are the length and overlapped interval of ships i and j, respectively, and γ_{ij} is the amount of those two ships' overlapping penalty that is obtained from the multiplication of two recent overlapped quantities in Cartesian time and place space. Using the power of 0.5 in the above equation is according to the experience, obtained from different performances. Maximum feasible error of a solution is accrued when two ships overlap, or

$$\gamma_{ij}^{max} = (A_{ij}' \times B_{ij}')^{0.5},$$

where

$$A_{ij}' := \frac{L_i + L_j}{2}, \text{and } B_{ij}' := \frac{C_i + C_j}{2},$$

are respectively the maximum length and overlapped interval of ships i and j, and γ_{ij}^{max} is the maximum overlapping penalty of these ships. The value of penalty γ_{ij} can be normalize by dividing the sum of γ_{ij} to the sum of γ_{ij}^{max} in the interval [0,1].

Moreover, the objective of the RBAP is to minimise the cost of handling all vessels and at the same time, to maximise the time buffer between the vessels. The cost of handling each vessel can be computed by summing up the berthing time of each vessel and its processing time and subtract the departure time or we can find the tardiness of each vessel can be computed by subtracting the finishing time from the expected departure time. The objective function used by the GA in MATLAB is the same objective function as that of the mathematical model. Thus, the value of Z can be computed:

$$Z = \sum_{v=1}^{V} C_{1v}(T_v + h_v - d_v)^+ + \sum_{v=1}^{V} C_{2v}|P_v - b_v|.$$

It is clear that the variation in Z is in the same direction of Z, but more mild and if occurred in the interval $[0, +\infty)$. By using this fraction function characteristic that $0 < 1/Z \le 1$ for each $Z \ge 0$, the objective function Z can also be normalized in the interval $(0,1]$ as $1/Z$. So, the proposed fitness function for this problem can be defined as follows:

$$Fitness = Z^{-1} - \sum \gamma_{ij} / \sum \gamma_{ij}^{max}.$$

The above fitness function for each produced solution consists of two parts; at the first part corresponds to the objective function value and the second part corresponds to the impossibility of that response to be penalised. The first part of the fitness function causes that the fitness be increased by reducing the objective function value, and the second part causes that the fitness be decreased by increasing the deviations from the restrictions of non-overlapping in time and place. As these two parts are normalized respectively in the intervals $[0,1]$ and $(0,1]$, the changes of fitness function will be in the interval of $(-1, 1]$.

4.4   Generating the Next Population

From the current population and after the fitness function values of its individuals are computed, a new population is generated using the genetic operators of crossover, mutation, and reproduction. These operators are described below.

Crossover Operator. Also called the recombination operator, it is one of the operators of GA used to generate the next population from the present one. It is applied to the chromosomes of a randomly selected couple of individuals (parents). The recombining of their genes results in a couple of new chromosomes (children). The way the recombination is implemented is as follows. Two random integers are chosen from interval $[0, 2|V|]$, where $2|V|$ is the length of a chromosome. These two integers point to two shear points in a chromosome, one low and one high. The genes to the left of the low shear point and those to the right of the high shear point of the first parent are copied into the chromosome of the first child. In the same way, using the same shear points, the chromosome of the second parent contributes to the chromosome of the second child. The genes between the shear points are generated using a decimal random number $\lambda \in [0,1]$ as follows,

$$Ch1 = [\lambda \times Par1 + (1 - \lambda) \times Par2], \text{ and}$$
$$Ch2 = [\lambda \times Par2 + (1 - \lambda) \times Par1]$$

where Par1 and Par2 are the middle sections of the first parent and the second parent, respectively, and Ch1 and Ch2 are the corresponding middle sections of the first and second child, respectively. It is clear that if Par1 and Par2 belong to a convex possibility set related to constraints (5), (6), and (15), Ch1 and Ch2 will be in this set too. This is because if x is the linear combination of two integer numbers y and z, $z \ge y$, and $y \le |x| \le z - 1$, then $y \le |x| + 1 \le z$. Equations (5), (6), and (15) are then checked for feasibility to confirm the new chromosomes. For illustration purposes, please see Figs. 6 and 7.

Parent1	7	12	25	50	33	70	20	95
Offspring1	7	12	25	49	38	72	28	95
Parent2	9	15	22	48	62	80	60	84

Fig. 6. Offspring1

Parent1	7	12	25	50	33	70	20	95
Offspring2	9	15	22	48	56	78	52	84
Parent2	9	15	22	48	62	80	60	84

Fig. 7. Offspring2

Mutation Operator. Mutation is another operator of GA which contributes to the generation of the next population. One of its characteristics is that it helps prevent the search process from getting trapped in a local optimum. To implement it, a chromosome is randomly selected from the current population. One of its genes is also randomly selected, then changed (mutated). A new chromosome is thus generated as a result. The mutation presented in this paper is as follows. If the selected gene is related to the berthing position, P_v (defined in Sect. 3.4), a random integer is selected in the interval $[0.5L_v, W - 0.5L_v]$ to replace the gene. If it is related to the berthing time, T_v (also defined above), a random integer is selected in the interval $[A_v, LN]$, where LN is a large number, to replace it. The new generated chromosome will satisfy conditions (5), (6), (7) and (10). The figure below illustrates the mutation operator (Fig. 8).

7	12	25	50	33	70	20	95
7	12	25	50	33	70	50	95

Fig. 8. Mutation operator

4.5   Stopping Criterion

GA stops when the maximum number of generations is reached.

5   Computational Experiments

Twenty instances of the mathematical model of RBAP with different numbers of vessels have been solved using B&C and GA. All instances have randomly generated arrival times, expected processing times, departure times, lengths of vessels and preferred berth places.

As mentioned earlier, B&C can only solve exactly small scale instances of RBAP. GA, on the other hand, provides is optimal or near optimal solutions for all instances. However, the quality of the approximate solutions may not be good enough. Therefore a hybrid meta-heuristic which combines both B&C and GA has also been implemented. The hybridisation scheme is a coarse hybrid heuristic which calls first B&C as implemented in CPLEX for short periods of time. This repeated call provides a number of good solutions which are then used to initialise the population of GA. Around 10 seconds of CPU time is allocated to B&C. The execution time of the hybrid is the aggregated execution times of B&C and GA.

GA, coded in Matlab, managed to solve all 20 instances. For the small size instances 1–5 and 11–15, the GA parameters of population size, probability of crossover, probability of mutation, and the maximum number of generations are 100, 0.8, 0.1, and 1000, respectively. In the case of the large size instances, 6–10 and 16–20, population size, probability of crossover, probability of mutation, and the maximum number of generations are set to 200, 0.8, 0.2, and 1500, respectively.

All 3 approaches B&C, GA and B&C+GA have been run on a PC with Intel Core 2 and 2.40 GHz CPU with 4 GByte RAM running Windows 7 Operating System. Note that CPLEX runs on instances 8–10 and 18–20 have been aborted after one hour of execution time. All computational results are recorded in tables Tables 2 and 3. These are self explanatory.

5.1 Test Results

In the instances considered, the number of constraints, the number of decision variables, and the B&C computational time grow exponentially with the

Table 2. Computational results for instances with $\lambda=5$

No	No.V	CPLEX		GA				B&C+GA			
		Obj	CPU time	Best Obj.	Av.Obj	Std	CPU time	Best Obj.	Av.Obj	Std	CPU time
1	5	51	1:00	61	88	3	2:16	–	–	–	–
2	6	22	1:00	37	49	2	2:45	–	–	–	–
3	7	87	1:00	115	149	8	2:41	–	–	–	–
4	8	91	1:00	132	169	5	2:56	–	–	–	–
5	9	144	5:01	183	300	15	2:40	145	156	6	2:54
6	10	119	15:29	176	221	29	4:37	119	119	0	4:45
7	15	306	36:52	432	544	31	4:51	337	351	16	5:16
8	20	152	1:00:00	263	423	41	6:19	228	253	15	6:27
9	25	257	1:00:00	425	522	38	6:53	370	445	19	7:08
10	30	438	1:00:00	614	851	45	7:28	539	722	16	7:43

Table 3. Computational results for instances with $\lambda=10$

No	No.V	CPLEX		GA				B&C+GA			
		Obj	CPU time	Best Obj.	Av.Obj	Std	CPU time	Best Obj.	Av.Obj	Std	CPU time
11	5	58	1:00	64	76	9	2:14	–	–	–	–
12	6	37	1:00	42	52	3	2:32	–	–	–	–
13	7	102	1:00	116	141	2	2:21	–	–	–	–
14	8	110	1:00	157	249	10	2:46	–	–	–	–
15	9	163	6:24	209	243	18	2:32	169	183	7	2:51
16	10	136	13:18	191	236	26	4:49	136	136	0	4:58
17	15	366	38:04	473	564	33	5:19	388	430	14	5:37
18	20	239	1:00:00	424	471	39	5:25	372	404	19	5:43
19	25	399	1:00:00	508	637	42	5:57	433	473	18	6:12
20	30	659	1:00:00	860	1024	37	6:32	748	861	23	6:51

increasing in the number of vessels. However, the CPU time required by GA and B&C+GA does not grow fast with the increase in the problem size. The average objective function and the standard deviation of GA and B&C+GA are recorded in the Tables 2 and 3. B&C managed to solve only few of the large size instances. GA, however, found the optimal or near solutions for all instances in reasonable CPU times (see columns 8 and 12 of Tables 2 and 3). Its effectiveness appears when in conjunction with B&C (hybrid meta-heuristic). The solutions became closer to the optimum ones.

The gap between B&C and GA solutions increases with the increase in the size of the problem (number of vessels) which translates into a large search space. In other words, as the problem size increases, GA will find less and less accurate solutions unless the time increases substantially. CPLEX, on the other hand, may not even be able to solve the problem in acceptable times. There is therefore a tradeoff between the performances of both approaches which is captured by the gaps between the solutions returned by the two algorithms. Hybridisation reduces this gap.

6   Conclusion

Berth allocation is one of the most important operations in container terminals. Determining the optimal berthing time and the best berthing position for vessels arriving at container terminals is essential for the efficient running of the these terminals. Here, a new mathematical model of the mixed integer programming type that addresses robust berth allocation is proposed. Its solutions help mitigate the uncertainty in arrival times and handling times of vessels. Instances of this model have been solved with an exact method namely B&C as implemented in CPLEX, an approximate approach namely the genetic algorithm and a hybrid of both which benefits from the exact nature of the former and the efficiency of the latter. The numerical results show that the hybrid meta-heuristic B&C+GA is superior to both B&C and GA in that it finds solutions to all problems in acceptable time and accuracy. B&C+GA is characterised by its coarse hybridisation nature. Future work will consider a finer and seamless hybridisation to reduce time overheads and improve solution quality. It will also consider comparisons with other meta-heuristics as have been introduced recently [1,3,16,18–21].

References

1. Blum, C.: Ant colony optimization: introduction and recent trends. Phys. Life Rev. **2**(4), 353–373 (2005). Elsevier
2. Cheong, C., Tan, K., Liu, D., Lin, C.: Multi-objective and prioritized berth allocation in container ports. Ann. Oper. Res. **180**(1), 63–103 (2010). Springer
3. Fister, I., Fister, J., Yang, X., Brest, J.: A comprehensive review of firefly algorithms. Swarm Evol. Comput. **13**, 34–46 (2013). Elsevier
4. Ganji, S., Babazadeh, A., Arabshahi, N.: Analysis of the continuous berth allocation problem in container ports using a genetic algorithm. Mar. Sci. Technol. **15**(4), 408–416 (2010). Springer

5. Goh, K., Lim, A.: Combining various algorithms to solve the ship berthing problem. In: 12th IEEE International Conference on ICTAI, pp. 370–375. IEEE (2000)

6. Guan, Y., Cheung, R.: The berth allocation problem: models and solution methods. OR Spectrum **26**(1), 75–92 (2004). Springer

7. Holland, J.: The Grid: Adaptation in Natural and Artificial Systems: An Introductory Analysis with Applications to Biology, Control, and Artificial Intelligence. University of Michigan Press, Ann Arbor (1975)

8. Imai, A., Sun, X., Nishimura, E., Papadimitriou, S.: Berth allocation in a container port: using a continuous location space approach. Transp. Res. Part B Methodol. **39**(3), 199–221 (2005). Elsevier

9. Lee, D., Chen, J., Cao, J.: The continuous berth allocation problem: a greedy randomized adaptive search solution. Transp. Res. Part E Logist. Transport. Rev. **46**(6), 1017–1029 (2010). Elsevier

10. Lee, Y., Chen, C.: An optimization heuristic for the berth scheduling problem. Eur. J. Oper. Res. **196**(2), 500–508 (2009). Elsevier

11. Li, C., Cai, X., Lee, C.: Scheduling with multiple-job-on-one-processor pattern. IIE Trans. **30**(5), 433–445 (1998). Springer

12. Lim, A.: The berth planning problem. Oper. Res. Lett. **22**(2), 105–110 (1998). Elsevier

13. Moon, K.: A mathematical model and a heuristic algorithm for berth planning. Brain Korea. **21** (2000). Citeseer

14. Moorthy, R., Teo, C.: Berth management in container terminal: the template design problem. OR Spectrum **28**(4), 495–518 (2006). Springer

15. Kaveshgar, N., Huynh, N., Rahimian, S.: An efficient genetic algorithm for solving the quay crane scheduling problem. Expert Syst. Appl. **39**(18), 13108–13117 (2012). Elsevier

16. Kennedy, J.: Particle swarm optimization In: Sammut, C., Webb, G.I. (eds.) Encyclopedia of Machine Learning, pp. 760–766. Springer, New York (2011)

17. Kim, K., Moon, K.: Berth scheduling by simulated annealing. Transp. Res. Part B Methodol. **37**(6), 541–560 (2003). Elsevier

18. Vazquez-Rodríguez, J.A., Salhi, A.: Hybrid evolutionary methods for the solutionof complex scheduling problems. In: Advances in Artificial Intelligence, pp. 17–28 (2006)

19. Salhi, A., Fraga, E.S.: Nature-inspired optimisation approaches and the new plant propagation algorithm. In: Proceedings of the ICeMATH 2011, pp. K2-1-K2-8 (2011)

20. Vazquez-Rodríguez, J.A., Salhi, A.: A synergy exploiting evolutionary approach to complex scheduling problems. In: Computer Aided Methods in Optimal Design and Operations, Series on Computers and Operations Research, pp. 59–68. World Scientific (2006)

21. Salhi, A., Vazquez-Rodríguez, J.A.: Tailoring hyper-heuristics to specific instances of a scheduling problem using affinity and competence functions. Memetic Comput. **6**(2), 77–84 (2014). Springer

22. Wang, F., Lim, A.: A stochastic beam search for the berth allocation problem. Dec. Supp. Syst. **42**(4), 2186–2196 (2007). Elsevier

23. Xu, Y., Chen, Q., Quan, X.: Robust berth scheduling with uncertain vessel delay and handling time. Ann. Oper. Res. **192**(1), 123–140 (2012). Springer

24. Zhen, L., Chang, D.: A bi-objective model for robust berth allocation scheduling. Comput. Ind. Eng. **63**(1), 262–273 (2012). Elsevier

25. Zhen, L., Lee, L., Chew, E.: A decision model for berth allocation under uncertainty. Eur. J. Oper. Res. **212**(1), 54–68 (2011). Elsevier

Dealing with the Strategic Level of Decisions Related to Automated Transit Networks: A Hybrid Heuristic Approach

Olfa Chebbi[1(✉)] and Jouhaina Chaouachi[2]

[1] Institut Supérieur de Gestion de Tunis, Université de Tunis,
41, Rue de la Liberté - Bouchoucha, 2000 Bardo, Tunisia
olfaa.chebbi@gmail.com
[2] Institut des Hautes Etudes Commerciales de Carthage, Université de Carthage,
IHEC Carthage Présidence, 2016 Tunis, Tunisia
siala.jouhaina@gmail.com

Abstract. The automated transit networks (ATN) is a new and sophisticated concept which has the possibility to solve problems related to transit in urban areas. In ATN, driverless vehicles run on exclusive guideways in order to provide on-demand transportation service. In this paper, we focus on the strategic level of decision related to ATN. We deal with the problem of determining the best size of fleet of ATN vehicles while satisfying a set of transportation demands. A hybrid heuristic approach is developed while taking into account the objective of finding good quality solutions in a short computational time. Computational results performed in this study demonstrate the efficiency of our approach.

Keywords: Automated transit networks · Driverless vehicles · Routing · Heuristic

1 Introduction

In this paper, we tackle the problem of defining the best fleet size option for automated transit networks (ATN). ATN is an on demand transportation system adapted for the person movements in urban areas. It offers a taxi-like service where vehicle move one person or a group of person of their choice.

In order to manage ATN, three levels of transportation decisions should be taken:

1. The strategic level of decisions which treats the decisions that have a long-lasting effect on the system under consideration. This includes decisions related to the number of stations, depots, the size of the stations, the network design, etc.
2. The tactical level of decisions: treat the decisions that are updated every week, month, or quarter. These decisions include purchasing decisions, scheduling decisions and transportation strategies.

© Springer International Publishing Switzerland 2016
M.J. Blesa et al. (Eds.): HM 2016, LNCS 9668, pp. 202–217, 2016.
DOI: 10.1007/978-3-319-39636-1_15

3. The operational level of decisions: refers to day-to-day decisions such as routing, decisions, optimization of the energy use, transit time, etc.

While ATN is known to be a sustainable and alternative solution for the use of private vehicles, it has not yet achieved a great success. In fact, studies show that driverless ATN vehicles are expensive [3]. Therefore, the tactical and strategic decision of purchasing ATN vehicles contribute on the total investment made to build an ATN system. Studies shows that purchasing driverless ATN vehicles contribute on 10 % on the total investment made for such a system [15]. Hence, the reduction of the cost of purchasing ATN vehicles is critical to reduce the ATN investment costs.

The limitations associated with the huge investment cost on purchasing ATN vehicles could be reduced through appropriate fleet sizing optimization procedure. In fact a well-designed ATN solution with an effective fleet sizing procedure would improve the efficiency of the system from the economic and financial perspective. Consequently, this paper focus on the fleet sizing of an ATN system.

In fact, we deal with a specific routing problem related to the strategic level of decisions. Our aim is to find the best size of ATN fleet in order to satisfy a set of transportation requests in a static deterministic setting. Earlier works on fleet sizing for ATN was proposed by Li Jie et al. [15] without considering battery issues of ATN vehicles. Battery constraints involve that a small set of the ATN fleet of vehicles would not be available for service as they are charging their battery. That is why, optimizing of the fleet size of an ATN system is of a high importance for such an intelligent system.

Genetic algorithm uses techniques inspired by natural evolution. By doing so, genetic algorithm allows a population of individuals to solve any combinato rial optimization problem. Genetic algorithms (GA) [22] have been implemented successfully to solve many routing problems [4]. However for more complicated and real world problems, GA needs to be coupled with problem-specific techniques in order to enhance its performance. Traditional heuristic methods are incorporated to enhance the performance of genetic search.

To solve the proposed ATN fleet size problem, we propose a hybrid two-stages GA. The first stage (initialization phase) consists of a linear programming heuristic inspired from the works in the literature [17] that generates a feasible solution to our problem. In the second stage (evolutionary phase), a GA procedure is applied while incorporating the obtained solution from the first phase. The proposed GA uses specific evolutionary techniques in order to search for the optimal solution or at least a near-optimal solution for the proposed problem. The whole scheme can be considered as a hybrid heuristic algorithm since it combines the linear programming heuristic as an initial solution generator and the GA meta-heuristic paradigms.

The contributions of this paper are several. In this paper we:

1. Present a routing problem related to ATN which consists on serving a set of predefined passenger's request. The objective of our problem is to determine the best feet size by reducing the number of used vehicle.
2. Present a hybrid heuristic algorithm based on GA to obtain good quality solution **in a reasonable computational time**.

The remainder of this paper is organized as follows: Sect. 2 presents the background of the paper. In Sect. 3, problem definition of minimizing fleet size of ATN and notation of model are presented. In Sect. 4, the proposed heuristic to deal with our problem is presented. Section 5 presents our computational results. Finally, the conclusions are drawn in Sect. 6.

2 Background of the Paper

Automated Transit Networks (ATN) is an on-demand transit system in which vehicles are designed to transport a small set of passengers on demand without any stops or transfer [1]. ATN uses exclusive guideways in order to combine the advantages of using rail transit with the benefit of private vehicles. ATN has the ability to provide a taxi like service. In fact in ATN, passengers are served only on-demand. In a specific ATN station, a passenger or a group of passengers come and ask to be carried to a specific ATN destination. An ATN vehicle is then automatically dispatched to them in order to take them to their final destination [29]. Stations in an ATN system are placed off the main line. Therefore, vehicles could accelerate/decelerate without interfering with other vehicles passing through the main line. That is why, an ATN system could offer a non-stop transportation service as vehicles could bypass all intermediate stations.

Currently, ATN vehicles are supported by specific modern technologies [23] that are usually designed to make the system run on electricity. The ATN system is fully automated which means that it is operated by computer control. Hence, ATN vehicle are driverless [6] and required no human intervention. ATN vehicles are also small which can accommodate up to six passengers.

ATN vehicles run on exclusive guideways. The guideways are designed in order to eliminate any interference with other transportation modes. Thus, ATN would reduce congestion on urban roads and contribute on increasing the sustainability of urban areas. ATN have been implemented in many real case applications such as in Heathrow Airport [13], Korea [24], Sweden [25] and United Arab Emirates [19].

2.1 Literature Review

Several works related to ATN have been proposed in the literature. We could note for example operational planning [7], empty vehicle movements [14], simulation [9,10], optimal design [27–29], energy minimization [8,11,18] and so on.

As for fleet size we could note the work of Li et al. [15], and Chebbi and Chaouachi [3,5] where it was proposed to develop a simulated annealing approach and mathematical formulation for the fleet sizing problem of ATN.

3 Problem Definition

In this section, we present the problem definition related to our problem as an extension to the ATN works of Mrad and Hidri [18]. In [18], Mrad and Hidri studied the problem of energy minimization of ATN in a static deterministic context. We extend in this paper their works by adapting the proposed problem to the context of fleet sizing of ATN systems. The ATN problem studied in this paper is a public transportation problem that aims to serve a set of ATN users transportation request under battery and time windows constraints.

Let us suppose to have an ATN system with a specified ATN network of a fully connected guideways. The objective of our problem is to find the number of ATN vehicles which will be deployed in the ATN network.

Let us suppose to have M station and one depot DP. In what follows, we suppose that we have a list of ATN transportation requests T. Let define Sp as a matrix cost of shortest time path between each two physical stations in the ATN network. Each transportation request i is characterized by departure station Ds_i, departure time Dt_i, arrival time At_i and arrival station As_i. An ATN vehicle could serve only one transportation request at a time. This is due to the fact that ATN system offers a taxi-like transportation service where passengers requests are not mixed within the same vehicle [13]. Our ATN routing problem focuses on minimizing the fleet size to satisfy all the available trips without any delay with respect to the battery and time windows constraints. In addition, we suppose that ATN pods(vehicles) need to visit the depot whenever it is necessary to load their batteries. The ATN routing problem can be described as a graph based problem where nodes represent passenger requests and edges represent movements of ATN vehicles between trips. In fact, it is more convenient to think about vehicles moving between transportation requests, rather than moving between physical ATN stations. Therefore, we can model trips as the nodes in a graph. We should note that modeling trips as a node in a graph based modeling of ATN is already proven to be a valid a viable approach and was used extensively in the literature (see for instances Lees Miller (2011) [13], Lees-Miller and Wilson (2012) [14]).

Consequently, our problem definition could be described briefly as follows. Let $G = (V, E)$ be a directed graph where $V = \{v_0, v_1, ..., v_n\}$ is a nodes set, node v_0 denotes a depot at which an unlimited number of identical ATN pods are based. The number of used ATN vehicles will be minimized and used as our objective function. The remaining vertices of V represent ATN user travel requests. $E = \{(v_i, v_j) : i \neq j\}$ is an arc set. Each arc (v_i, v_j) has an associated nonnegative electric consumed energy c_{ij} and a nonnegative travel time t_{ij}. The ATN fleet sizing problem consists of designing a set of least cost vehicle routes such that: (i) Every route starts and ends at the depot. (ii)Every trip is visited exactly once by exactly one vehicle. (iii) The total consumed electric energy of any route can not exceed the battery capacity B. (iv) Each trip has a specific departure time, arrival time and departure station. If the ATN vehicles arrive at a station to satisfy a trip x before the departure time of x, the ATN vehicle should wait at its current location until the trip x is trigged. (v) No waiting time is allowed for transportation requests. Consequently, a transportation request i should be served at exactly its departure time Dt_i.

We define also $V^* = V \setminus v_0$

The set of edges E will be defined by the following rules:

– For each nodes v_i and v_j in V^*, we add an arc $(i; j)$ only if: $j > i$ and $At_i + Sp(As_i; Ds_j) < Dt_j$. This condition satisfies the time window constraints for the reason that is not possible to serve a trip after its departure time Dt_i. The cost of this arc is noted by c_{ij} and t_{ij}. c_{ij} represents the energy consumed from arrival station of trip i (As_i) to arrival station of trip j (As_j) by passing the departure station of trip j. In the same way, $t_{ij} = Sp(As_i; Ds_j) + Sp(Ds_j; As_j)$.

– For each node v_i we add an arc $(0; i)$ (the cost of this arc is c_{0i} and t_{0i}. c_{0i} represents the energy used from the depot to the arrival station of trip i while passing the departure station of trip i. In the same way, $t_{ij} = Sp(DP; Ds_i) + Sp(Ds_i; As_i)$.

– For each node v_i, we add an arc $(i; 0)$. The cost of this arc is c_{i0} and t_{i0}. c_{i0} represents the energy used from the arrival station of trip i to the depot. $t_{ij} = Sp(As_i; DP)$.

We denote also $E^* = \{(c_i, c_j) : i \neq j \text{ and } i, j \in V^*\}$.

From this problem definition, we can see that our problem is related to the asymmetric distance-constrained vehicle routing problem (ADCVRP) [26]. The distance constraint is imposed by the battery capacity in this case. Our problem is NP-hard, and is asymmetric because the distance from node i to node j is different to the distance from node j to node i.

3.1 An Assignment-Based Formulation

In this section an integer programm which is based on an assignment formulation [18], is presented. To that aim the following decision variables and definitions are introduced.

– $x_{ij} = 1$, if node j is visited immediately after node i, and 0, otherwise.
– z_i is the amount of charge used to reach the node $i \in V^*$ from depot.
– $a_i = c_{0i}$ for $i \in V^*$.
– $b_i = B - c_{i0}$ for $i \in V^*$.

Hence, the minimum charge assuring the trips is equal to the optimal value of the following programming model.

$$\mathbf{ATN(1)}: \quad \text{Minimize} \sum_{(i) \in V^*} x_{0i} \tag{1}$$

$$\sum_{j \in \delta^+(i)} x_{ij} = 1 \forall i \in V^* \tag{2}$$

$$\sum_{j \in \delta^-(i)} x_{ji} = 1 \forall i \in V^* \tag{3}$$

$$z_i + c_{ij} \leq z_j + (b_i - a_j + c_{ij})(1 - x_{ij}) \quad \forall (i, j) \in E^* \tag{4}$$

$$a_i \leq z_i \leq b_i \quad \forall i \in V^* \tag{5}$$

$$x_{ij} \in \{0,1\} \forall \ (i,j) \in E \tag{6}$$

$$z_i \geq 0 \ \forall i \in V^* \tag{7}$$

The objective (1) is to minimize the number of used vehicles to satisfy the set of transportation requests T. Constraints (2 and 3) require that each node $i \in V^*$ must be visited and left only one time, respectively. Constraints (4) ensure the following conditions.

- if $x_{ij} = 1$, then $z_i + c_{ij} \leq z_j$, with $a_i \leq z_i \leq b_i \ \forall i, j \in V^*$

Constraints (5) present a trivial bound limitation on the charge needed to perform the trip i. Finally, Constraints (6) and (7)indicates that the decision variables x_{ij} are binary-valued and z_i positive real variable.

4 Hybrid Genetic Algorithm for ATN Fleet Sizing

This section describes our hybrid heuristic based on the combination of linear programming heuristic and a GA. The first phase in our algorithm is the initialization phase in which a feasible solution is obtained using a relaxed linear program. The second phase consists of an adapted GA to obtain good quality solution of our problem.

In this section, the GA is first described. Then, the first phase of our hybrid genetic algorithm is explained and detailed.

4.1 First Phase Procedure

In this section, we present a linear programming method which will be used as an first phase procedure for our algorithm. In fact, we consider solving the mathematical formulation presented in Sect. 3 while relaxing constraints 4, 5 and 7.

However by solving this linear program, we can get infeasible roads starting and ending at the depot which consume more energy than what the Battery allows. To fix this dilemma, we propose a specific algorithm in order to get feasible solution of the relaxed linear program. In fact starting from the set of obtained roads, we correct the infeasible ones by considering the trips forming these roads as an input for a new smaller problem which could be solved to optimality using the mathematical model proposed in Sect. 2. An illustration of our first phase method is proposed in Fig. 1. Based on the assumption that we have 10 ATN transportation requests, the obtained roads of the relaxed linear program are illustrated in Fig. 1. We could note that we obtained 2 infeasible roads. To get a feasible solution from the output of the relaxed linear program, we submit each infeasible road to the proposed mathematical model in Sect. 3. Consequently, we get a feasible solution which would be considered as the input for our second phase of the hybrid GA.

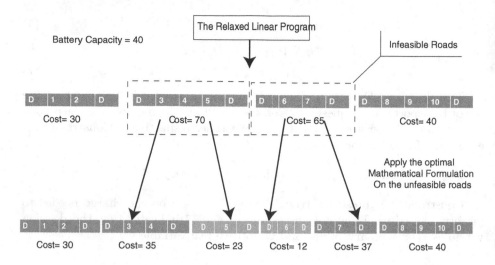

*D represents the Depot Node

Fig. 1. Illustration of initial phase procedure

4.2 Genetic Algorithm as a Second Phase for Our Hybrid Heuristic

In this section, we detail the main components of the GA which is used as a second phase in our hybrid heuristic.

Representation and Evaluation Function. The solution in our algorithm are represented as a permutation of travels (see Fig. 2). As an evaluation function, we used the split function of Prins proposed initially for the capacitated vehicle routing problem [21]. This function offers the advantage of generating feasible roads starting and ending at a specific node called the depot with respect to several constraints (the battery constraint in our case). This function builds an auxiliary graph where each node represents a trip in the solution plus a dummy node. Each edge represents a road that starts and ends in the depot while passing by one or several nodes. The shortest path in this auxiliary graph represents the best splitting option of a permutation. Figures 2, 3 and 4 present an illustrative example of the evaluation function [8]. Starting from an initial solution a, b, c, d, e, the auxiliary graph H is constructed as presented in Fig. 3. Based on H, a shortest time path is computed from the node 0 to the node 5. The roads related to the shortest time path (shown in bold edges) represent the final solution of the split function as shown in Fig. 4

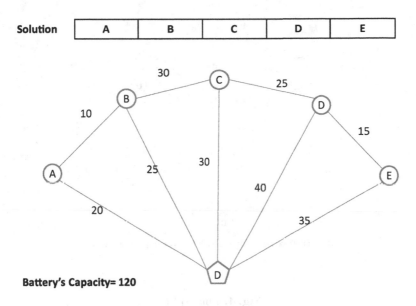

Fig. 2. Initial solution

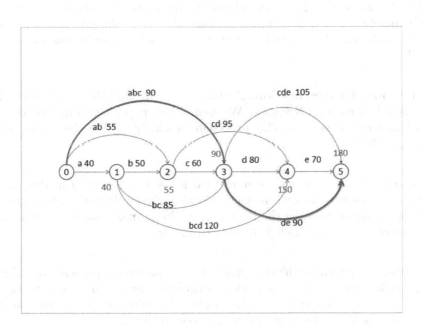

Fig. 3. Intermediate graph for the split function

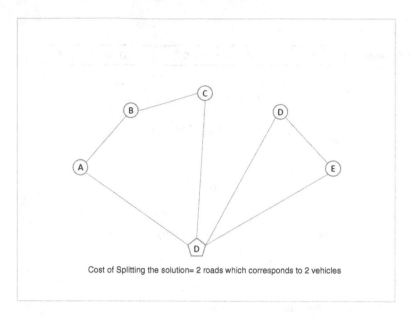

Cost of Splitting the solution= 2 roads which corresponds to 2 vehicles

Fig. 4. Final split

Initial Population. The first population is generated at random. The use of the We add also to the first set of individuals the solution obtained from the first phase of our hyper GA. Generating random solution would guarantee to have a diverse population which would guarantee to have a fast convergence rate of our algorithm.

Chromosome Representation. A chromosome denotes the sequence of trips to be visited by each ATN vehicle. We used in this paper a trip routing permutation based representation. This procedure is widely used in the literature for routing problems [16].

Crossover. A one-point crossover method is used and applied in this paper. This procedure is used as it ensure that the order of visit of trips in a permutation is at least swapped between the beginning of the permutation and a randomly chosen point. A brief example is illustrated in Fig. 5.

Mutation. We use the exchange mutation procedure in the proposed algorithm. This procedure ensures more variations than other mutation operators [12]. In the exchange mutation, two trips are selected at random and their positions are exchanged. An illustrative example is depicted in Fig. 5.

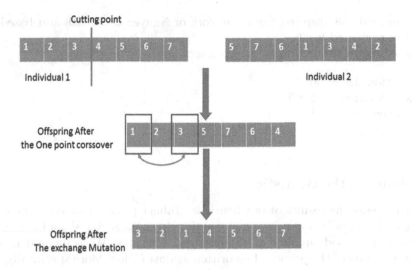

Fig. 5. One-point crossover and exchange operator.

5   Computational Results

In this section, we describe the experiments made to apply our algorithms to a set of problems generated randomly. For each instance, tests are made using a program coded in C++ and simulations are performed on a computer with a 3.2 GHZ CPU and 8 GB of RAM. All the mathematical models were solved using Cplex 12.2 commercial solver.

5.1   Test Problems

In this paper, we used the instances from the literature [3,5] based on the instance generator of [18]. We test our algorithm on 19 different size classes n where $n \in \{10; 15; 20; 25; 30; 35; 40; 45; 50; 55; 60; 65; 70; 75; 80; 85; 90; 95; 100\}$. For each size class, 40 instances were generated. So, in general we did tested the proposed heuristics on 760 different instances. For the evaluation of results we used the GAP metric which is defined as follow:

$$GAP = (\frac{S_{Heuristic} - LB}{LB}) * 100$$

- $S_{Heuristic}$ is the solution of the heuristic.
- LB is the minimum between our GA and the results of the mathematical formulation presented in this paper [3].

The different mathematical models presented in this paper were developed using Cplex 12.2. To calibrate the algorithm and find reasonable values for its different parameters, we conducted preliminary computational experiments using a set of instances with size equals to 50 transportation requests. The parameter tuning method was inspired from the work of Nguyen et al. [20] and based on the works presented in [2].

Consequently, the hybrid GA parameters were sets as follows:

- Population size = 50
- Max generations = 2000
- Crossover rate = 0.9
- Mutation rate = 0.3

5.2   Result of the Heuristic

Table 1 presents the results of our heuristic. Table 1 presents for each class size the average Gap in % and computational time in seconds. We should note from our results the goods quality of the obtained solutions. We compared in this paper the results of the proposed algorithm against Cplex. More specifically, the Cplex uses an exact method to solve the proposed problem. Consequently, it would be interesting to compare the performance of our hybrid heuristic against

Table 1. Result of the hybrid GA

Size of the problem	Average Gap%	Average time sec
10	19.583	1.829
15	13.071	1.948
20	21.750	2.313
25	24.113	2.652
30	13.709	3.005
35	17.890	3.393
40	12.885	3.812
45	11.340	4.190
50	10.495	4.665
55	5.987	5.293
60	7.946	5.724
65	8.573	6.280
70	3.063	6.924
75	6.374	7.525
80	4.440	8.253
85	2.705	8.785
90	2.548	9.482
95	1.805	9.933
100	0.228	10.528
Average	9.921	5.607

an exact method. In fact, the results of Cplex [3] deteriorate as the size of problem increases. This results on the fact that the average GAP for the hybrid GA is decreasing which proves that it founds better results than Cplex in many large size instances. Also, our method is fast as it founds its solution in an average time equal to 5.607 s. Especially for large instances our method found feasible solution in less than 11 s. Therefore, we should note that our approach is successful as our primary objective was to find feasible solution in a small computational time.

5.3   Analysis of the Results

In this section, we wanted to compare our results against those of Cplex for solving the mathematical model presented in Sect. 2. For that purpose, we run the mathematical model on the same instances used to test our heuristic. We set the maximum time resolution of Cplex to 100 s. A first comparison between the two proposed approach is given in Table 2 where we present a set of descriptive statistics of the obtained fleet size.

Table 2. Descriptive Statistics of the obtained results expressed in term of Fleet size

	hybrid GA	Cplex
Minimum	2	2
25 % Percentile	7	8
Median	13	14
75 % Percentile	18,75	19
Maximum	31	32
Mean	13.14	13.73
Std. Deviation	6.388	6.779
Std. Error of Mean	0.2317	0.2459

Next, we wanted to use an enhanced statistical analysis method to know whatever the results are significantly different or not. For that purpose, we performed first a normality test to know whatever the obtained results are derived from a normal distribution. These results confirm that our two series of data are not derived from normality distribution[1]. Therefore and in order to compare the obtained results, we used the Wilcoxon matched-pairs signed rank test.

Results of this test are presented in Table 3. These results confirm that the results are statistically different and that the proposed heuristic is better than the results of Cplex. This is enhanced for the case of large instances (> 80).

Finally, we wanted to analyze the running time of our hybrid heuristic. First, we propose to run a correlation test in order to study the relation between the size of the problem and the running time of the hybrid GA. We used the Pearson correlation test and its results confirm the statistical relation between the two variables[2].

---

[1] For the three normality tests, we found a P-value <0.0001.
[2] For the Pearson correlation test, we found a P-value <0.0001 in addition to an r statistic equals to 0.9929.

Table 3. Results Of the Wilcoxon matched-pairs signed rank test

Statistic	Value
P value	<0.0001
Significantly different? (P <0.05)	Yes
One- or two-tailed P value?	Two-tailed
Sum of positive, negative ranks	37707 . −112171
Sum of signed ranks (W)	−74464

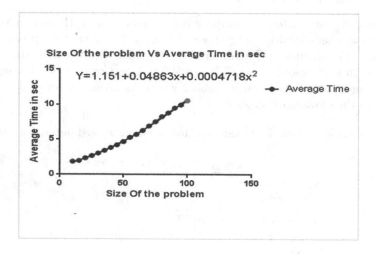

Fig. 6. Average time as a function of the Size Of the problem

Table 4. Results of the post non linear regression analysis Runs test

Statistic	Value
Points above curve	8
Points below curve	11
Number of runs	8
P value (runs test)	0.1994
Deviation from Model	Not Significant

Next, we wanted to derive a non linear regression equation in order to estimate properly the running time as a function of the size of the problem. For that purpose. We used a second order polynomial (quadratic) equation to estimate the running time as a function of the problem size. Results are presented in Fig. 6. We used also a Runs test as a post analysis step of our equation.

Results of the Runs test are presented in Table 4 and confirm that our equation estimates properly the running time of our heuristic as a function of the size of the problem.

6   Conclusion

In this paper, we considered the offline routing problem of the ATN that uses electrical vehicle with batteries that can be recharged only in the depot in order to find the best size of the fleet of vehicles. This problem was solved by a hybrid heuristic which was coupled with a optimal mathematical model adapted to the characteristics of our problem.

The proposed algorithm was tested on a large set of 760 generated instances of the ATN with 1 depot, 12 stations and a battery that make the vehicle run for 40 min. Taking into account the difficulty of our problem, our method is competitive since it uses the solution given by the Cplex solver. We found and average Gap of 9.921 % in 5.607 s. Our Next step is to design tight and sophisticated hybrid heuristic to tackle the stochastic fleet size optimization problem related to ATN.

References

1. Anderson, J.E.: A review of the state of the art of personal rapid transit. J. Adv. Transp. **34**(1), 3–29 (2000)
2. Chebbi, O., Chaouachi, J.: Effective parameter tuning for genetic algorithm to solve a real world transportation problem. In: 2015 20th International Conference on Methods and Models in Automation and Robotics (MMAR), pp. 370 375. IEEE (2015)
3. Chebbi, O., Chaouachi, J.: Optimal fleet sizing of personal rapid transit system. In: Saeed, K., Homenda, W. (eds.) CISIM 2015. LNCS, pp. 327–338. Springer, Heidelberg (2015)
4. Chebbi, O., Chaouachi, J.: Reducing the wasted transportation capacity of personal rapid transit systems: An integrated model and multi-objective optimization approach. Transportation Research Part E: Logistics and Transportation Review (2015)
5. Chebbi, O., Chaouachi, J.: Simulated annealing approach for solving the fleet sizing problem in on-demand transit system. In: Abraham, A., Wegrzyn-Wolska, K., Hassanien, A.E., Snasel, V., Alimi, A.M. (eds.) AECIA 2015. AISC, pp. 217–226. Springer, Heidelberg (2016)
6. Choi, K., Jang, W.: Development of a transit network from a street map database with spatial analysis and dynamic segmentation. Transp. Res. Part C Emerg. Technol. **8**(1–6), 129–146 (2000). http://www.sciencedirect.com/science/article/pii/S0968090X00000139
7. Fatnassi, E., Chaouachi, J., Klibi, W.: Planning and operating a shared goods and passengers on-demand rapid transit system for sustainable city-logistics. Transp. Res. Part B Methodol. **81**, 440–460 (2015)
8. Fatnassi, E., Chebbi, O., Chaouachi, J.: Discrete honeybee mating optimization algorithm for the routing of battery-operated automated guidance electric vehicles in personal rapid transit systems. Swarm and Evolutionary Computation (2015)
9. Fatnassi, E., Chebbi, O., Siala, J.C.: Evaluation of different vehicle management strategies for the personal rapid transit system. In: 2013 5th International Conference on Modeling, Simulation and Applied Optimization (ICMSAO), pp. 1–5. IEEE (2013)

10. Fatnassi, E., Chebbi, O., Siala, J.C.: Two strategies for real time empty vehicle redistribution for the personal rapid transit system. In: 2013 16th International IEEE Conference on Intelligent Transportation Systems-(ITSC), pp. 1888–1893. IEEE (2013)

11. Fatnassi, E., Chebbi, O., Siala, J.C.: Comparison of two mathematical formulations for the offline routing of personal rapid transit system vehicles. In: The International Conference on Methods and Models in Automation and Robotics (2014)

12. Jun, S., Park, J.: A hybrid genetic algorithm for the hybrid flow shop scheduling problem with nighttime work and simultaneous work constraints: A case study from the transformer industry. Expert Syst. Appl. **42**, 6196–6204 (2015)

13. Lees-Miller, J.D.: Empty vehicle redistribution for personal rapid transit. Ph.D. thesis, Liverpool John Moores University (2011)

14. Lees-Miller, J.D., Wilson, R.E.: Proactive empty vehicle redistribution for personal rapid transit and taxis. Transp. Plan. Technol. **35**(1), 17–30 (2012)

15. Li, J., Chen, Y.S., Li, H., Andreasson, I., van Zuylen, H.: Optimizing the fleet size of a Personal Rapid Transit system: A case study in port of Rotterdam. In: International Conference on Intelligent Transportation, pp. 301–305 (2010)

16. Matei, O., Pop, P.C., Sas, J.L., Chira, C.: An improved immigration memetic algorithm for solving the heterogeneous fixed fleet vehicle routing problem. Neurocomputing **150**(Part A), 58–66 (2015). bioinspired and knowledge based techniques and applications The Vitality of Pattern Recognition and Image Analysis Data Stream Classification and Big Data Analytics Selected papers from the 16th International Conference on Knowledge-Based and Intelligent Information; Engineering Systems (KES2012) Selected papers from the 6th Iberian Conference on Pattern Recognition and Image Analysis (IbPRIA2013). http://www.sciencedirect.com/science/article/pii/S0925231214012387

17. Mrad, M., Chebbi, O., Labidi, M., Louly, M.: Synchronous routing for personal rapid transit pods. J. Appl. Math. **2014** (2014). http://dx.doi.org/10.1155/2014/623849

18. Mrad, M., Hidri, L.: Optimal consumed electric energy while sequencing vehicle trips in a personal rapid transit transportation system. Comput. Ind. Eng. **79**, 1–9 (2015). http://dx.doi.org/10.1016/j.cie.2014.09.002

19. Mueller, K., Sgouridis, S.P.: Simulation-based analysis of personal rapid transit systems: service and energy performance assessment of the masdar city prt case. J. Adv. Transp. **45**(4), 252–270 (2011). http://dx.doi.org/10.1002/atr.158

20. Nguyen, V.P., Prins, C., Prodhon, C.: A multi-start iterated local search with tabu list and path relinking for the two-echelon location-routing problem. Eng. Appl. Artif. Intell. **25**(1), 56–71 (2012)

21. Prins, C.: A simple and effective evolutionary algorithm for the vehicle routing problem. Comput. Oper. Res. **31**(12), 1985–2002 (2004)

22. Qi, Y., Hou, Z., Li, H., Huang, J., Li, X.: A decomposition based memetic algorithm for multi-objective vehicle routing problem with time windows. Comput. Oper. Res. **62**, 61–77 (2015). http://www.sciencedirect.com/science/article/pii/S0305054815000891

23. Soulas, C.: Automated guideway transit systems and personal rapid transit systems. In: Papageorgiou, M. (ed.) Concise Encyclopedia of Traffic & Transportation Systems, pp. 36–49. Pergamon, Amsterdam (1991). http://www.sciencedirect.com/science/article/pii/B9780080362038500158

24. Suh, S.D.: Korean apm projects: status and prospects. In: Proceedings of the 8th International Conference on Automated People Movers, San Francisco, CA (2001)

25. Tegnér, G., et al.: PRT in sweden: from feasibility studies to public awareness (2007)
26. Toth, P., Vigo, D.: The Vehicle Routing Problem. Monographs on Discrete Mathematics and Applications, Society for Industrial and Applied Mathematics (2002). http://books.google.co.uk/books?id=TeMgA5S74skC
27. Won, J.M., Lee, K.M., Lee, J.S., Karray, F.: Guideway network design of personal rapid transit system: A multiobjective genetic algorithm approach. In: 2006 IEEE Congress on Evolutionary Computation, vols. 1–6 (2006)
28. Won, J.M., Choe, H., Karray, F.: Optimal design of personal rapid transit. In: Intelligent Transportation Systems Conference, pp. 1489–1494, September 2006
29. Zheng, H., Peeta, S.: Network design for personal rapid transit under transit-oriented development. Transp. Res. Part C Emer. Technol. (2015). http://www.sciencedirect.com/science/article/pii/S0968090X15000674

DEEPSAM: A Hybrid Evolutionary Algorithm for the Prediction of Biomolecules Structure

Moshe Goldstein[1,2(✉)]

[1] Computer Science Department, Jerusalem College of Technology, 91160 Jerusalem, Israel
goldmosh@g.jct.ac.il
[2] The Fritz Haber Research Center for Molecular Dynamics, Institute of Chemistry,
The Hebrew University of Jerusalem, 91904 Jerusalem, Israel
goldmosh@fh.huji.ac.il

Abstract. *DEEPSAM* (*D*iffusion *E*quation *E*volutionary *P*rogramming *S*imulated *A*nnealing *M*ethod), a hybrid evolutionary algorithm, is presented here. This algorithm has been designed for finding the global minimum, and other low-lying minima, of the potential energy surface (PES) of biological molecules. It hybridizes Evolutionary Programming (EP) with two well-known global optimization methods (the Diffusion Equation Method - DEM, and a kind of Simulated Annealing - SA), and with the L-BFGS quasi-Newton local minimization procedure. This combination has produced a powerful tool (a) for finding a good approximation of the native structure of a protein or peptide, given a Force Field (FF) parameters set and a starting (unfolded) structure, and (b) for finding an ensemble of structures close enough structurally and energetically to the native structure. The results obtained until now show that DEEPSAM is a powerful structure predictor, when a reliable FF parameters set is available. DEEPSAM's implementation is time-efficient, and requires modest computational resources.

Keywords: Hybrid evolutionary algorithm · Evolutionary programming · Diffusion equation method · PES smoothing · Simulated annealing · Protein structure prediction

1 Motivation

It is well known that the folded state of a peptide (or a protein) is the conformation with the most thermodynamic stability. The global minimum of its force field (FF) – of its Potential Energy Surface (PES) – is one among an exponential set of conformations that are PES local minima. Unlike techniques which mostly rely on structural information of known peptides and proteins, the algorithm presented here relies strictly on the PES. In spite of their doubtful accuracy, the standard empirical FFs are, at present, the available tools for molecular PES modeling. Assuming the correctness of such a FF (at $T \rightarrow 0$ K), we were interested in an algorithm to look for the global minimum of the PES. The global minimization algorithm presented here, called DEEPSAM [1] (Diffusion Equation Evolutionary Programming Simulated Annealing Method), has been developed to reach that goal.

© Springer International Publishing Switzerland 2016
M.J. Blesa et al. (Eds.): HM 2016, LNCS 9668, pp. 218–221, 2016.
DOI: 10.1007/978-3-319-39636-1_16

2 Design Overview

The approach taken in DEEPSAM assumes that important progress can be made by constructing a *hybrid* algorithm that combines several well-established optimization techniques of complementary advantages, allowing good global exploration and good local exploitation of the topography of the PES. Following this approach, DEEPSAM was designed as a PES global minimization *hybrid evolutionary algorithm* (EA), built upon the TINKER molecular modeling package.

The population-oriented approach of Evolutionary Programming (EP) was adopted in DEEPSAM, allowing a parallel search of the PES. In DEEPSAM, like in most EAs, the size N of the population is constant during all the run. At its *initial population generation* step [2], DEEPSAM starts by randomly choosing an ensemble of N structurally dissimilar and physically feasible local minima conformations. This determines N widely distributed sub-regions of the PES, which are simultaneously explored, in parallel.

In order to overcome energy barriers found in every one of those PES sub-regions, specially designed *mutation operators* called DEMSA (*D*iffusion *E*quation *M*ethod with *S*imulated *A*nnealing), were designed. DEMSA operators are PES function transformation operators. They are adaptable combinations of PES smoothing (provided by the Diffusion Equation Method (DEM)), Simulated Annealing (SA) and the L-BFGS quasi-Newton local minimization procedure. By smoothing the PES, fewer minima have to be sampled, reducing problem size and search effort, and allowing energy barriers at the unsmoothed PES to be overcome. Instead of randomly changing coordinates, DEEPSAM uses a Levy-distribution-based method to probabilistically choose PES smoothing levels upon which specific DEMSA operators are generated on-the-fly. Those DEMSA operators are a kind of meta-mutation operators which are created as dynamically-chosen combinations of SA with itself and/or with L-BFGS, one applied upon the smoothed PES and the other one applied upon the unsmoothed PES. The actual PES smoothing is chosen from a range of smoothing levels that adapts itself to the computation conditions, by accordingly extending or constraining itself. Offspring conformations of a given parent conformation are generated by applying those dynamically generated DEMSA operators, in parallel.

A specially designed *survivor selection operator* is used in order to choose the conformations that will be part of the next population. For each one of the N parent conformations in the current population: (a) five DEMSA operators are probabilistically generated as was explained above, (b) a "family" of five offspring conformations is generated by applying those five DEMSA operators upon the parent conformation. Hence, at each iteration, DEEPSAM generates N families of six conformations, each. For each one of those N families, the offspring conformations and their parent compete among themselves. The survivor conformations are selected by using a Metropolis-like criterion. The set of N selected conformations becomes the new population. For each survivor conformation, the algorithm adapts itself by deciding which type of DEMSA operator to use in the next iteration. Each one of the DEMSA operators to be actually applied upon the conformations of the new population is selected depending on energetic and geometric considerations.

Additionally, DEEPSAM keeps a population of *"best-so-far" local minima*, in which the best conformations found during the run are kept. If the best-so-far population as a whole does not improve, the algorithm is re-initialized in order to try to improve population diversity.

In summary, DEEPSAM dynamically changes itself in accordance to the evolution of the computation.

DEEPSAM stops according to an external parameter which determines how many iterations it will run until it stops. Most of the results reported in our already published work, were obtained after an average of 100 iterations over populations of size five (a small population size indeed). At the end of the run, when the number of iterations is reached, the energetically deepest conformation among those in the best-so-far population is assumed to be the global minimum. It is significant to note that DEEPSAM ends not only with the deepest minimum found, but with an ensemble of deep lying minima structures.

3   Implementation Overview

DEEPSAM is built in two layers: (a) the upper layer (written in Python) is the actual DEEPSAM's implementation, which was designed with two levels of parallelism - at each iteration, N independent sets of five independent DEMSA operators are applied in parallel upon the N conformations of the current population; (b) the lower layer, upon which DEEPSAM is built, is a set of TINKER's programs (written in Fortran77) that are used by the Python code as operators, applied as necessary upon molecular conformations represented by TINKER-xyz (and TINKER-seq) files, in accordance with the corresponding TINKER-prm FF parameters file.

4   Work in Progress

At this moment, several research projects are in progress, in which biomolecules structure prediction is needed, and DEEPSAM is being used for this purpose. As soon as results will be available, they will be published.

An embedded language for Python, called EFL (Embedded Flexible Language) [5, 6] has been developed by the author and colleagues, at the FlexComp Lab (http://flex-comp.jct.ac.il) of the Jerusalem College of Technology. A rewriting of DEEPSAM is planned, using this new parallel programming methodology and embedded language. Also, in collaboration with Prof. Miroslav Popovic (from Novi-Sad University, Serbia), a version of DEEPSAM is planned, which will use Software Transactional Memory (STM).

5   Concluding Remarks

DEEPSAM has been successfully applied to the prediction of the native structure of neutral cyclic peptides. Those calculations were done in the gas phase and with implicit solvent models, producing results in good agreement with experimental data [1, 3]. Using DEEPSAM, a detailed full atomistic geometry of the Ubiquitin +13 ion (in mass spectrometric conditions) has been predicted [4]. Also, structure prediction calculations were done for several +6 charge distributions over Ubiquitin - the results will be published in the near future.

Our already published work shows that DEEPSAM is an effective structure predictor which has a good run-time performance, made possible by its parallel implementation, its self-adaptability, and the small population size used (five conformations per population, in most of the cases), which means relatively small computing resources requirements.

References

1. Goldstein, M., Fredj, E., Gerber, R.B.: A new hybrid algorithm for finding the lowest minima of potential surfaces: approach and applications to peptides. J. Comput. Chem. **32**, 1785–1800 (2011)
2. Fredj, E., Goldstein, M.: A knowledge-based approach to initial population generation in evolutionary algorithms: application to the protein structure prediction problem. In: Dershowitz, N., Nissan, E. (eds.) Choueka Festschrift 2014, Part I. LNCS, vol. 8001, pp. 252–262. Springer, Heidelberg (2014)
3. Goldtzvik, Y., Goldstein, M., Gerber, R.B.: On the crystallographic accuracy of structure prediction by implicit water models: tests for cyclic peptides. Chem. Phys. **415**, 168–172 (2013)
4. Goldstein, M., Zmiri, L., Segev, E., Wyttenbach, T., Gerber, R.B.: An atomistic structure of ubiquitin +13 relevant in mass spectrometry: theoretical prediction and comparison with experimental cross sections. Int. J. Mass Spectrom. **367**, 10–15 (2014)
5. Yehezkael, R.B., Goldstein, M., Dayan, M., Mizrahi, S.: Flexible algorithms: enabling well-defined order-independent execution with an imperative programming style. In: Proceedings of ECBS-EERC 2015, pp 75–82. IEEE Press (2015)
6. Dayan, D., Goldstein, M., Popovic, M., Mizrahi, S., Rabin, M., Berlovitz, D., Berlovitz, O., Bussani, Levy, E., Naaman, M., Nagar, M., Soudry, D., Yehezkael, R.B.: EFL: Implementing and testing an embedded language which provides safe and efficient parallel execution. In: Proceeding of ECBS-EERC 2015, pp. 83–90. IEEE Press (2015)

Author Index

Printed in the United States
by the publisher

Printed in the United States
By Bookmasters